Dr. med. Werner Hendrik Cornelis

Are we too many?

2nd edition
(November 2014)

An essay

Environment
Overpopulation
Remedies

Copyright 2014 W. H. Cornelis, author
All rights reserved
ISBN 978-1-291-79400-7

A timely essay

This essay intends to be a message, not a divertimento.

The starting-points are the problems of environment.
Nine threats for environment are being defined, and hundreds of data, numbers and formula are presented.

For this reason alone, this booklet is a valuable document.

These data constitute the foundation for conclusions that will bring about a revolution in our way of thinking, in our planning of the future.
Hence, it seemed appropriate to pay attention to this extensive list of data.

We are not the first to concern ourselves with the problem of overpopulation.

Thomas Robert Malthus was the first to raise questions about food shortage and population numbers in his book "*An essay on the Principles of population*" (1798)

Paul Ralph Ehrlich, in his book "*The Population bomb*" (1968) was especially afraid of the finiteness of the fossil reserves.

The club of Rome voiced Ehrlich's ideas. The Club played and still plays an important role.

The Kyoto protocol (1997) is an official agreement targeting the greenhouse gases and the global warming but ignoring the population numbers.

This essay makes the connection between the 9 threats for the environment on the one hand, and the overpopulation on the other.

For the first time an estimate is being made on the quantity of the overpopulation which is rated at about 70% if the individual consumption remains unchanged; but still about 40 %, even when the individual consumption is halved.

Finally, solutions are being presented through the tackling of the 9 threats and manipulating the population numbers through migration and birth ratio. Special attention is focused on the Chinese way of handling this problem.

This essay, as the first, makes that important fifth step.

Most of all, it constitutes an essential step towards durable peace between all the men and all the peoples of Our World.

Table of Contents

Chapter 1: **Historical data**

Prehistory
Eighteenth century
Malthus
Ehrlich
Club of Rome
Kyoto

Chapter 2: **The nine main challenges for mankind and environment**

Facts, data, statistics, interferences
How serious are these threats?
What can we do about them?

Chapter 3: **The tenth threat: population pressure**
How can we quantify?

Historically, today's numbers
Geographic distribution
Urbanisation
Influence of the level of Education
How to measure the population pressure?
What is an ecological footprint?
What is a population index?
What does the Kyoto-protocol propose?

Chapter 4: **Population pressure: +70 %: how to deal with it?**

The four angles of Incidence
Immigration and emigration
Birth Rate and mortality, birth rate policy

Chapter 5: **Our leverages**

Social and fiscal measures
How to convince those concerned?
Moral considerations

Appendix: **Water, the most important material on earth**

Chapter 1

Historical data

A million years of mankind

Around 1800: the questions arise

Thomas Robert Malthus

Paul Ralph Ehrlich

The Club of Rome

The way to Kyoto

A million years of mankind

About a million years ago, somewhere in Africa, apes descended from a tree. They called themselves humans. They lived in caves and caverns, where they found better protection. The woods were still their nearby store-room.

Those woods were used as firewood. Animals were caught and fruits harvested. This area expanded, the distance to their cave became bigger, their habitat larger, thus they needed huts on their trek.

When the area was exhausted, the humans tried their luck elsewhere, in a new hut; they became nomads. This way, they found new means of existence and in the meantime the original area could recover.

When the original area had recovered, it could be re-used. This system could be repeated and so it was, for thousands of centuries. There was plenty of room for the humans in those days.

It all changed when man discovered agriculture, some 12000 years ago. It put an end to the nomadic existence. Man didn't have to move any longer: he could grow his own vegetables; he could harvest and keep domesticated animals within reach.
Civilization could arise, with settlements, villages, towns, hierarchy, writings, rules and laws. Soon some kind of social security based on solidarity emerged, and schools, markets, trades, duties and functions arose. All this created space for many more people.

There was plenty of space; so nobody had questions.
Yet in some places negative effects became apparent: deforestation and desertification are occurring. But: who cares? the space available seemed limitless. We'll start elsewhere…

Not before the early 19th century people started questioning and looking for answers.
Nowadays the environmental issue has become one of the most threatening problems for humanity. Our sheer existence is at stake. New facts, questions and challenges make us look for new solutions. Really new solutions!

Albeit that they may interfere with our self-sufficiency and our complacent way of living.

At the end of the 18th century,
people started asking questions.

Those were real pioneers,
not always appreciated by their contemporaries.

For a long time they were the only ones
to look for answers to questions that have become
commonplace in the last 30-40 years only.

Today there are thousands of publications on this subject that concerns everybody;
a subject that is truly of vital importance.

Everybody will agree.

The ARAL lake, Central Asia, is empty due to *homo sapiens*.
Photo: Staecker, via Wikipedia

Thomas Robert Malthus

Born 13-02-1766, Surrey, England.
Died 23-12-1834, Bath, England.
Since 1790: clergy.

Malthus' field: macroeconomics, demography and evolutionary economics.
He studied philosophy and mathematics in Cambridge.

Between 1798 and 1826 he published 6 versions of what would become his life work

"An Essay on the Principle of Population".

Every edition contained new material and refuted the critics; yet his ideas evolved as well.

The basis of his idea:

> *The population grows logarithmically.*
> *The food supply increases arithmetically.*
> *Starvation must be the consequence. (Disastrous undersupply)*

Yet he didn't disapprove of an increase in population *per se*.

When he wrote this essay, the industrial revolution hadn't occurred. He could not foresee the increase in agricultural productivity. Even less could he predict the way in which the industrial activity was going to jeopardise life itself.

Malthus was a pioneer.

But he only discussed **one** of the environmental problems that menace us to day:
Food shortage.

For ages people have vilified him, among others by Christian dignitaries.

Malthus will be back.

Paul Ralph Ehrlich
(°1932 Philadelphia, USA)

Ehrlich originally was a biologist and entomologist specialised in butterflies, but he astonished the world with his 1968 book "**The population bomb**", in which he predicted that the human population growth in the 70's and 80's would lead to enormous problems.

Sooner or later the natural resources would be extinct.
This was also the motor of research done by the Club of Rome.

Their predictions were similar.

The crisis did not occur, but when early in this millennium
oil prices started to explode, a period of reflection occurred.

How to Know the Butterflies (1960)
Population, Resources, Environments: Issues in Human Ecology (1970)
Introductory Biology (1973)
The End of Affluence (1975)
Extinction (1981)
The Cassandra Conference: Resources and the Human Predicament (1988)
The Birder's Handbook: A field Guide to the Natural History of North American Birds
(1988, with David S. Dobkin and Darryl Wheye)
The Population Explosion (1990)
Human Natures: Genes, Cultures, and the Human Prospect (2002)
One with Nineveh: Politics, Consumption, and the Human Future
(2004, co-authored with Ann Ehrlich)

The Club of Rome - 1968
(source: Wikipedia)

The Club of Rome is a private foundation (1968) without political or economic power, founded by the Italian industrialist Aurelio Peccei and the Scottish scientist Alexander King. The club meets every year in a different country.

The executive committee consists of 14 members.

In 2007 there were 78 active members and 36 associated members.
It has 53 honorary members, among them Queen Beatrix of Holland
and prince Philip of Belgium.

Its objectives are:

- The inter-relation of the global problems (population growth, food production, industrialization
  extinction of natural resources, pollution, etc.) elaborated in a world model.
- To show the world the graveness of the problems: the growth of the economy and of the world population, and the increasing gap between rich and poor
- To urge governments and politicians to take coordinated measures

The Club of Rome produced several reports. The best known is **"The Limits to Growth"** (1972)

This book is in alignment with Paul Ehrlich's book "The population bomb" and makes a connection between economical growth and the impact on environment.

Malthus only spoke about food;
The Club of Rome also brought natural resources into the discussion.

Critics state that technology will solve all problems.
The predictions of the Club of Rome have not (yet) been realized.

The report has been translated in 37 languages and 12 million copies have been sold.
In 2004 an updated version has been published.

Kyoto (1997)

1988 Toronto, Canada:

First meeting on a high level by politicians and scientists about climate change.
The industrialized nations agree on a voluntary reduction by 20% by 2005.
(known as the Toronto objective)

1990 Sundsvall, Sweden:

Second meeting of IPCC (International Panel on Climate Change): The lack of absolute scientific certainty cannot be invoked as an excuse for inertia.

1990 New York, United States:

United Nations resolution 45/212 lays the foundation of UNFCCC in preparation of the world summit in Rio in 1992 (The United Nations Framework Convention on Climate Change). The UNFCCC wants to stabilize the concentration of greenhouse gases in the atmosphere to a level that will exclude dangerous human influences on the climate.

1991 New York, United States:

The CO_2 emission should be reduced to the level of 1990 by the year 2000.
George Bush Sr. prevents these resolutions to be enacted.

1992 Rio De Janeiro, Brazil:

World summit! All nations are invited to sign the protocol within a certain time frame.

1994 The AOSIS (Association of small insular states) urges for more severe measures

1995 Berlin, Germany:

Climate summit with the main world leaders: strong measures wanted!

1995 Italy:

2000 scientists declare that there is increasing evidence of the human influence on the climate change.

1996 Switzerland:

A trade in emission rights is being put forward.

1997 Belgium:

    The USA and Japan block any agreement.

1997 New York, United States:

    Disappointing intervention by US president Bill Clinton.

1997 KYOTO, Japan:

    168 countries, including 15 EU countries agree about gases: carbonic acid, methane, nitrogen oxide, CFK, SF_6,... The requirements are not on a high level: the emissions must be 5% lower by 2005 in comparison to the level of 1990.

    The USA refused to ratify.
    Russia signs, as the last country in October 2004.
    As a consequence of which the protocol is at last implemented worldwide on February 16, 2005.

2007: A reduction of 20% of emissions is put forward within the next 20 years.
Again, the objectives are very modest.

CONCLUSION:

After the problem of **food shortage** (Malthus, 1798),

After the problem of **fossil materials** (1968, Ehrlich, Club of Rome),

Kyoto puts the emphasis on **greenhouse gases,** and **global warming.**

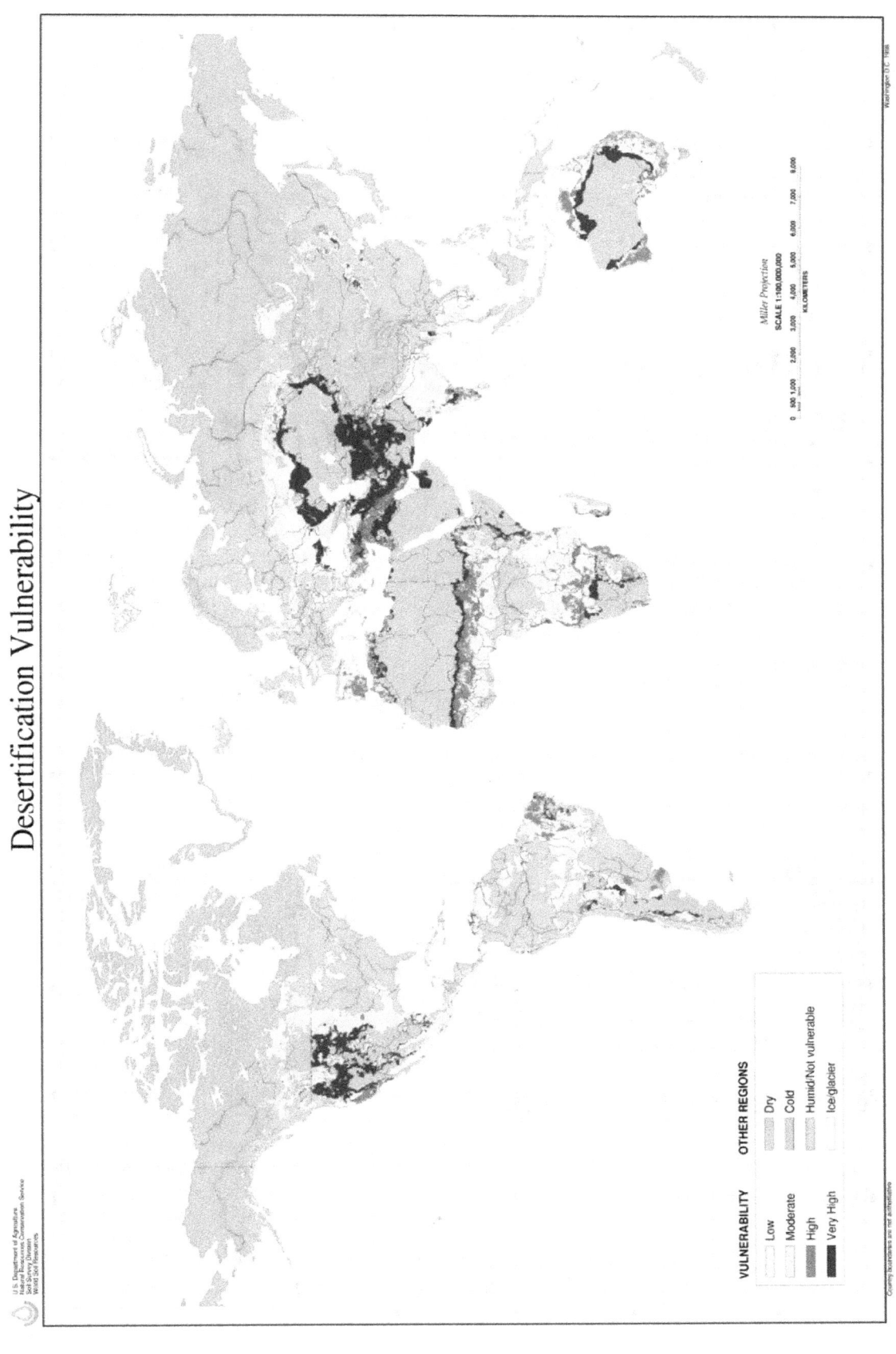

Source: U.S. Department of Agriculture Washington D.C. 1996

Chapter 2

The nine main challenges for mankind and environment

Up to now we encountered 4 threats for environment.
They constitute 4 threats to mankind as a whole.

In an historic sequence we mentioned:
food shortages, the exhaust of fossil reserves, the global warming and the greenhouse gases.

In recent years new threats have come forward.

The (provisional complete) list runs as follows:

1. Desertification
2. Deforestation
3. Food shortages
4. Overfishing
5. Exhaustion of the supplies: ores and minerals
6. Exhaustion of the supplies: energy bearers
7. Exhaustion of the supplies: water
8. Global warming
9. Greenhouse gases and the ozone hole

Many of these threats have an influence one on another; almost always they intensify each other's destructive action.

We will review them separately, point out their synergy and appraise their threat.
And above all we will address the main issue:

What can we do about it?

1. Desertification

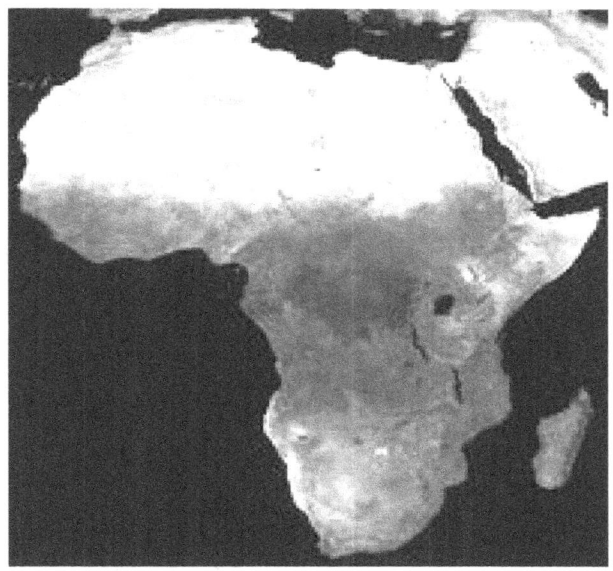

Africa seen from space

Most striking is the abundance of deserts, also in Arabia.

25 years ago, the "Horn of Africa" (Ethiopia and Somalia) consisted of 40% of forests, now it is down to 4%.

The remaining area turned into desert, not into farming land.

Have a close look at Madagascar: where has all the green gone to?

Some data:

Worldwide, every year, the deserts grow by 12.000.000 ha. = 120.000 km², i.e. 4 times the surface of Belgium. Every year 25 billion tons of farming soil is lost. Each year the Sahara grows by 5-10 km. in North-South direction. As a consequence, migratory birds can't get across any more.

The cause lies with the humans and their goats (other farming animals died long ago). The destruction of the fertile equatorial rainforests only produces arid deserts, e.g. in Madagascar, where we see totally barren plains where no crops can grow. Water alone is not the solution: organic material is needed as well: (manure, humus, compost, etc.)

It has been said that forests came before human beings, and deserts followed them.

35 % of the Earth surface is at risk of turning into desert: see map on page 13. 4,5 Billion dollars are needed preventively, *every year*, during the next 20 years, just to prevent the situation from getting worse. The desert is a porous soil; the little rainfall easily seeps through to deeper layers; one may presume, that there is abundant phreatic water present (and indeed there is…)

But the ground water layers are not replenished, as there is no rain!

Hence, they can only be used for durable and essential purposes, e.g. reforestation, not for human consumption, not for the animals. "Recent" deserts, like the Sahel, have fertile soil at a depth of 5 feet.

The surface of a desert consists almost exclusively of mineral, inorganic material. Reforestation requires water and organic material. Only in those circumstances growth is possible.

Only then can animals like moles, worms and insects pull twigs and leaves under the surface. Only then can microbes break everything down into nutrients thus preparing the soil for planting.

There is a point of no return where the erosion maintains itself at an ever-increasing rate and the natural process of re-vegetation cannot regenerate. The lack of water, the frost and the large fluctuation in temperature between day and night turn the soil into sand that blows away in the wind.

What can we do about it?

Preventive: Diminish the human pressure.
Pre-desert areas such as the Sahel are in fact inhabitable. There is no other solution.

On a **curative** level: this topic will be discussed in the chapters "deforestation" and "water"

"The difference between a desert and a garden is not the water, but man"
(Tunisian proverb)

2. Deforestation

The history of civilization is one of deforestation. In the old Mesopotamia, Ur, Uruk and Babylon were responsible for massive deforestation and desert creation.

Today, the increasing population pressure is still the cause for progressing deforestation. The rainforest is (was) 11 million km², about 1/7 of the earth's surface; in South America 57%, in Southeast Asia 24%, in Africa 19%.

Deforestation is caused by tree cutting, fire (often kindled) and acid rain.

Nowadays, things are evolving rapidly:

- In the last 40 years, half the rainforest has disappeared
- Every minute the surface of 7 soccer fields, 4 million acres a year since 1955
- In the last 20 years, an area the size of France has vanished
- About 2 billion m3 of wood is used every year
- The Chinese use about 45 Kg of paper and cardboard per person per year, Americans 300

- The astronauts saw that the smoke column in Amazonia (1988) was as big as South Africa
- Yet agriculture is a complete failure: the soil is too poor. The Brazilian government is aware of it, they still hand out tree cutting permits
- 25 years ago, Ethiopia counted 40% of forests, now only 4%. This did not lead to more farming, but to more desert
- What do you think will happen in Borneo and Sumatra, where tropical forest is replaced by palm plantations? We all now know what happens in (sub) tropical regions when the forest disappears

This is extremely hazardous, especially for biodiversity.

Every year, 27000 species disappear; when is it homo sapiens' turn?

Early life on earth undoubtedly was vegetable (algae), animal life came later. Plants absorbed the abundant CO_2 and produced O_2. Only afterwards animal life was possible.

This CO_2-O_2 cycle is as necessary as ever before, between animal and human (with his polluting machines) on the one hand and the forests on the other.

What can we do about it?

As stated previously, we will treat desertification and deforestation together.

Preventive:
Stop cutting the forests, especially in the tropics.
Avoid forest fires (arson) and acid rain.
Diminish the population pressure.

Curatively:
Plant new forests, lots of forests. At least 7 million km², the size of the USA, not counting Alaska; and especially in the tropical areas, e.g. in the actual deserts. In that area, the danger for increasing desertification is clearly the greatest. But also in the so-called temperate zone, e.g. Spain and Scotland, where the forests were used as firewood, and for building ships and houses.

It would not be fair though, to demand that **they** should keep their forests and **we** would keep our airports, industrial areas etc. Therefore we all have to replant forests, partly as natural parks, not only for forestry.

Reforestation of the desert is a labour- and energy intensive task, but examples in Israel, Ghana and China prove that it can be done. The Beijing-Tianjin project deals with 190.000 km² in ten years, which is 37% of the area threatened by desertification in China. Already more than 38.000 km² were saved from desertification. In this case, they turned imperilled agricultural soil, not desert, into forests.

In many deserts there is enough phreatic water to allow reforestation but not enough to maintain meadows and fields. Trees use less water and for a shorter period than crops. E.g. in the Sahel it is possible to use the fertile layer under 1,5 m through a special technique, known as "deep rooting" for the trees.

In order to obtain a local climate change we will have to plant thousands of kilometres of forests; tons of organic material must be brought in; the new plants have to be protected against rodents, against malevolent people, sandstorms etc. We have to choose the right kind of trees and plants and introduce the right mycorrhizae.

But it can be done.

Most of the new forests must be natural forests, primitive forests with minimal human presence. This should enhance the **biodiversity**. Other parts of the new forest can be used for forestry: cut and replanted on a regular base. In fact: recycled. Only when these forests are fully functioning, it may be possible to re-introduce agricultural activity in small parts of the forest.

Acid rain

Acid rain is an important element in the destruction of the forests. Hence, it is appropriate to deal with this subject under the heading of "deforestation". Acid rain not only harms forests, but also soils, lakes, rivers, even rocks, sculptures and buildings.

Some facts:

- We are talking about "acid rain", but 20-60 % is in fact "dry rain".
- This phenomenon was first mentioned in 1852 by Robert Angus Smith (Manchester) Harold Harvey, about 1960, found a "dead lake" in Canada.
- The normal acidity (pH) of rain is 5.6, ergo slightly acid (by the CO_2). We talk of acid rain if the pH is below 5.
- The lowest ever measured was a pH of 2.2! (like vinegar)
- Involved are SO_2 and N_xO_x gases, chemical substances originating from the burning of fossil, sulphur containing fuel.
- This results in a yearly SO_2 output of 70 Tg, the volcanoes produce 8 Tg, wildfires 2.8 Tg. (One Tg or Teragram equals 1 million tons).

Adverse effects:

- Souring of rivers and lakes
- Lakes and rivers die
- Fish die
- Roe doesn't mature at a pH lower than 5
- Insects die, as do trout
- Forests die, hence less photosynthesis, more CO_2, less oxygen
- Buildings and statues are damaged
- Rocks and buildings disintegrate to gypsum ($CaSO_4$) that flows away, causing erosion

What is the remedy against acid rain?

Diminish the emission of gases by power plants by using low sulphur fuels and through an improved filtering. On this domain, we can already see some improvement.

What about the cars? Low sulphur fuel? Only renewable energy?
Using less energy; is this achievable with an ever-increasing world population?

Recycling and regeneration are the key words for future.

We will hear about them very often!

3. Food shortage

Some facts:

- The world population increases by 80 million every year (the population of Germany or Mexico i.e. 200,000 daily), yet the food supply does not increase or at least at a much smaller rate.
- According to the World Bank 2.8 billion people live below the poverty limit earning less than 2 $ a day.
- 800 million people suffer malnutrition.
- Every day 46,000 people die from hunger or from diseases caused by hunger.

Conclusion:

The standard of living of half the world population is now lower than a century ago.
The crisis is here and now and it is here to stay.

1. Today, the production of grain could suffice for 10 billion vegetarians, but the grain is in part used for the cattle. In order to produce 1 kg. of meat, we need 4-6 kg. of grain.
It is obvious that there is no room on Earth for 10 billion omnivores.

2. At the end of the eighties, a team of Brown University estimated that the world could feed 5.5 billion vegetarians, but only 3.7 billion people living like the South Americans and only 2.8 billion people living the North American way. Americans leave behind a larger footprint.

3. In the wealthy North the use of biodiesel and bio-ethanol is being promoted. This means that food crops will be used to generate fuel. That also means that the already scarce food surface will be overburdened, that food prices will rise- which is happening now. During the last 8 years, food prices have risen by 75%, the price of wheat by 200%. (April 2008). Many farmers are anxious because their fields and storages are being looted. Three agents of the World Food Program were killed in Sudan while trying to distribute free food.

4. Rich people spend a relatively small part of their budget on food; poor people spend about 60 to 70% of their income on food; for them an increase of the food prices is lethal; they have nothing left for clothing, accommodation, education etc.

5. People are not taking this any longer, which explains the food protests in Egypt, Mexico, Haiti, Bolivia, Ivory Coast and Indonesia. 3,318 protests in 3 weeks in 2008. Armed soldiers had to intervene; as a result dozens of people were killed or wounded. In April 2008, Wall-mart and Costco introduced rationing of rice. Vietnam and other rice-producing countries stopped their export. Etc, etc,…

6. There are many ominous signs on the wall.

All minerals and fossil energy carriers (coal, oil, uranium,...) are present as reserves.
They are by definition restricted and finite. *There is no more.*

The same goes for water: its supply is limited. *There is no more.*

The same *does not* go for food: it is produced every day, over and over again.
Then again, the agricultural surface is limited, especially the water supply.

As a result of irrigation, fertilization and even genetic manipulation, production can be enhanced. But somewhere, somehow, production reaches its limits.

This is all there is.

In 1798, Thomas. R. Malthus called this **Disastrous Undersupply**.

And what does the rich world do? It *burns* food. They grow wheat in order to make bio-ethanol; they make palm oil plantations in order to make biodiesel.

Food becomes unaffordable for the poor. The number of starving people increases. Poor people can no longer meet their basic needs, while rich people make airplane trips and drive gas guzzling cars.

One full gas tank for an SUV uses enough wheat to feed a person for a whole year. Above all, in order to produce the bio diesel, rainforests are cut in Borneo and Sumatra.
(see deforestation)

What can we do?

Nothing, really…

Food production has reached its limit, mainly because of water shortage.

This means that we cannot increase food production. All reserves of our planet
are already in operation.

An increase of the demand can no longer be met by higher production; food is going to become scarce and hence even more expensive when crops are not used for the production of biodiesel or bio-ethanol.

This is detrimental for the have-nots since prices rise too high.

As a result we see food riots. They will not stop.

Again, we must conclude that the main cause of the problem is in the population pressure, the only element that we can influence.

4. Overfishing

Of the nine threats we are dealing with, this one is the most serious, causing the most negative implications and the one that will be the first to show its malignant effects.

Some facts:

1. The FAO concluded in 2007 that 70% of all fish species are totally exploited or even extinct. Fish vanishes from the ocean, 10 to 15 fishing grounds are almost exhausted.

2. In the developing nations fishermen haul empty nets because of massive and subsidized overfishing by the rich countries with their industrialized boats, which leave nothing for the locals, not even for the future, since they destroy spawning and breeding areas.

3. For years the ocean around Newfoundland was considered the storeroom of Europe and America, especially for cod. The fishing industry, amongst others with trawlers, has annihilated the fish stock; even young fish were lost, before they could reproduce. The breeding spaces and their habitat were destroyed. A moratorium was imposed on fishing in this area. Yet the fish stocks did not recover, not even tens of years later.

4. Between 1950 and 1994 the amount of fish caught increased by 400%; in 1989 the amount was 82 million tons.

5. Fish don't have reserves like coal, oil, uranium, ore; only the natural growth of the fish population can be considered as an eternally usable reserve. When fishing exceeds this amount, the regeneration of the fishing grounds is jeopardized, with exponentially lowering of the stocks and reserves in future. Many areas already have reached this stage.

6. Bear in mind that fish is the main source of proteins for 1/5 of mankind.

The magnitude of the problem of overfishing is often overlooked, given the competing claims of deforestation, desertification, energy resource exploitation and other biodiversity depletion dilemmas.

	King Crab Northeastern Pacific Depleted	**Koningskrab** Paralithodes Camtschaticus
	Atlantic Cod Northeastern Atlantic Depleted to Overfished	**Kabeljauw** Gadus Morhua
	Haddock Northwestern Atlantic Depleted to Overfished	**Schelvis** Melanogrammus aeglefinus
	Salmon Northeastern Pacific Overfished to Steady	**Zalm** Salmo Salar
	Silver Hake Northwestern Atlantic Abundant	**Wijting** Merluccius Bilinearis
	Bluefin Tuna Atlantic Depleted	**Blauwvin Tonijn** Thunnus Thynnus
	Shrimp East central Pacific Overfished to Steady	**Garnaal** o.a. Crangon Crangon
	Atlantic Redfish Northeastern Atlantic Overfished to Steady	**Roodbaars** Sebastes spp.
	Alaska Pollock Northeastern Pacific Overfished to Steady	**Alaska Pollack** Theragra Chalcogramma
	Pacific Halibut Northeastern Pacific Overfished to Steady	**Heilbot** Hippoglossus spp.
	Atlantic Mackerel Northeastern Atlantic Overfished to Steady	**Makreel** Somber Scombrus
	Albacore Tuna, yellow Fin Pacific Overfished to Steady	**Albacore, geelvin** Thunnus Alalunga

Current Status: DEPLETED OVERFISHED STEADY ABUNDANT
based on 1992 data from the United Nations Food and Agriculture Organization

More facts:

1. Most problems of overfishing started in the fifties with technological progress. Fishing used to be a matter of luck; nowadays radar and sonar can be used.

2. Industrialized ships remain on sea for weeks; they have on board every tool to kill and freeze and can their catch; supplies are provided by sc. "hunting ships". As a consequence, the catch increased from 1950 to 1970 by 7% per year. In the last 25 years, the best 20 fishing grounds have disappeared, all empty. Many others are on the brink of extinction.

3. Cod reproduces very rapidly and can normally survive heavy fishing. But a long period of overfishing and the destruction of the vegetation on the sea floor by trawlers completely rooted out the cod in the Atlantic.

4. Fishermen often use fine meshed nets; little fish are not treated as food but as manure.

5. Nets are totally indiscriminate: for each ton of shrimp caught, 3 tons of fish is killed; 20,000 porpoises die every year in the nets of fishermen going after salmon. Tens of thousands of dolphins die every year in the nets of tuna fishermen.

6. Keep in mind that the quantity of fish should not be regarded as a reserve. The natural growth of the fish stock is the only reserve. If we catch more, we diminish the breeding stock and endanger the future of our children and grandchildren and even our own.
It goes without saying that this is an urgent problem.

7. We can differentiate between
   - a biological overfishing where fish mortality reaches a level where the biomass decreases and ultimately disappears
   - economic or bio-economic overfishing where it is no longer profitable to try to catch fish

8. On the other hand, jellyfish - cockroaches of the sea - are appearing in areas where they were never seen before. This is a result of the global warming and of the fact that they have no natural enemies. Et cetera…

9. Overfishing is without any doubt the most urgent problem of all nine:
   - because of its impact
   - because we are too late already, lagging decades behind
   - and above all: the supply cannot be increased; ergo the demand has to decrease; as a consequence, the impact by the population has to decrease

On the one hand we established that the total amount of fish cannot be considered as a reserve; the only real reserve lies in the natural growth of the fish stocks. If we don't realize this, we will destroy all breeding stock and also the future of the next generation of fish and of mankind.
On the other hand the critical limit has already been crossed in many areas; we therefore have to give time and opportunity to the breeding stocks, also for the sake of biodiversity.

Fish "farms" only offer a marginal contribution, and they may locally cause serious ecological problems by monoculture, antibiotics, etc... It is obvious that in the next few years fish production will not increase; on the contrary, it will have to come down.

Let us also be aware of the social corrections we will have to introduce in order to provide fish and food for the poor people, especially in the developing world.

On a preventive scale we must use less destructive fishing techniques that respect the stocks and breeding grounds. We cannot influence the amount of fish; the only variable we can control is the amount of people who are living from fish i.e. diminish the population pressure.

On a curative level the same measures apply, namely let the breeding stocks regenerate. Only in that way we can guarantee recycling, THE keyword regarding the environment.

*One evening, in a pond, the two last fish (m. and f.) are the only ones to restore the fish population.
How long, you think, is their life expectancy?*

Answer: see below

*Fishermen in Kochi, India
photo: Wikipedia*

Answer:

*Till the next dawn.
You surely guessed.*

5. Ores and minerals

Some facts:

In the case of ores and minerals we can truly speak of reserves. Energy carriers (oil, coal, gas, uranium) present themselves as reserves as well, but they merit a different approach, because energy can be produced aside of the reserves, and soon maybe instead of these reserves (see next contribution).

The reserves of ore and minerals do not grow.
They are used, and they are finite. *One day they will be gone.*

Hence, recycling is, once again, mandatory.
Only what is recycled, will be ready for new utilisation later.

And this is, in this case, quite feasible.

It does not look as if ore and minerals will be a problem, now or later, precisely because they are so easily re-used. New mines might be necessary though, causing further deforestation, or soiling of the earth by mercury, cyanide et cetera.
In this respect overpopulation might be a problem after all.

6. Energy

Some considerations:

1. The earth's crust provides us with energy carriers, a reserve that is naturally finite and not expandable. **Today, 90% of our energy comes from those fossil carriers.** Taking in account the present day consumption, we can rely on a reserve of natural gas for the next 60 years, of oil for the next 40 years, and of coal for 125 years.

2. According the IEA (International Energy Agency) the demand will double by 2030. The estimations put forward above may well be too optimistic.

3. The increased demand is caused by population growth, by increased consumption in Asia and the higher standards of life worldwide. **The biological footprint keeps growing**!

4. Fossil fuels must therefore be replaced by "alternative" energy, which is practically inexhaustible: biomass, water-, wind- and solar energy, but also geothermic energy, waves and tides can be considered. All these procedures are mainly meant to generate electricity.

5. According to the European Commission, 20% of the electricity must be generated by alternative sources by 2020. For Belgium e.g. the goal is 13%, but in 2008 the achievement is 2.2% only. World wide 1% of electricity results from wind energy, 0.1% from solar energy. Flanders obtains 3% of electricity from "green sources", mainly biomass.

6. The cost per KWh from turbines on land is *on the whole* comparable to the cost of electricity from coal– or nuclear plants. The KW from turbines in sea is 50% more expensive; the KWh-price from solar cells is about 5 times higher.

7. Many things still need to be done to improve the efficiency of the photovoltaic cells; furthermore: the silicon is still quite expensive.

8. The use of crops in order to produce bio-diesel or bio-ethanol has been discussed in a previous chapter (food shortage). Morally and socially it seems quite unacceptable. Besides: more energy can be produced on the same area by photo-cells, than by crops.

9. Energy carriers do contribute to the production of greenhouse gases, and to global warming. We will deal with these items later.

What can we do about this?

Our energy carriers (coal, gas, oil,...) are fossil remnants of earlier vegetable and animal life. They mostly contain combinations of carbon and hydrogen. For a long time, they were used as basis for coal-chemistry and petrochemical industry. These raw substances were the basis for making paints, plastics, and medicaments. All of them are very valuable substances.

As a first reflexion: let us first make these products, recycle them a number of times, and burn them eventually, only after their useful cycle. Mohammad Rezā Shāh Pahlavi of Persia found, 50 years ago, that petroleum was too valuable just to be burnt.

Since reserves are finite by definition, and since the most optimistic prognosis foresees their extinction in a few decades, we must NOW prepare a society moved by renewable energy only.

This means, that we will lack the 90% of today's energy. This, in turn, means that we have to replace them by other, more perpetual sources - or that we have to use much less energy. This is feasible, but only if we change our living pattern drastically, or if we try to lower the number of consumers.

At any rate heavy investments and serious research will be needed in order to make that transition, and a profound mental adjustment. Biodiesel and bio-ethanol, grown on soil that today provides food, is not acceptable for moral and ethical reasons. We must find other solutions: consume less, and with less consumers.

Fossil sources (coal, gas, oil) are too valuable just to be burnt and converted into useless and even harmful products such as CO_2 and water. We should convert them into useful products such as medicines, paints and plastics. Today we are wasting these valuable substances.

The next generation will condemn us severely for our arrogant way of life, that didn't take into account the future.

Smog in Beijing (right); the same location after 2 days rain (left) (august 2005, Wikipedia)

7. Water

Some materials are essential to life.
Oxygen, for instance, although some living creatures can do without it.
Some creatures can live without chlorophyll, others without haemoglobin.

No creature however, plant or animal, can live without water.
Water is absolutely indispensable for life.
There is no substitute, no replacement

The amount of water on earth has been the same since the beginning of times:
water is not being created; the same water circulates and is being recycled
for the last 4,5 billion years: *there is no new water.*

The global amount of water available on this planet equals about 1.359 billion km^3, of which only 0.6% is available as fresh water. Water has several unique physical and chemical qualities. We must assure its regulating and buffering role, which makes life on this planet possible.
(see appendix)

Water consumption		
	For toilet.	50 litre.
	Bath/ shower	30 litre.
In Western Europe: 140 litre/person/day	Dishes, laundry	35 litre.
In New York: 1,000 litre/person/day	Cleaning	10 litre.
	Varia: car, garden	9 litre.
	Food, drinking	6 litre.

According to the WWF though, we use 40 times more "virtual water" for the production of food, fabrics etc. An average Englishman uses 4,645 liters of water every day! The same goes for an average Western European.

Do keep in mind that our "water footprint" mainly lies abroad, but that doesn't alter the problem. Agriculture consumes more than 60% of the world water. It is easier to reforest the desert than to turn it into agricultural soil. The USA squanders the most water, using 20% of it, with only 4% of the world population.

According to the World Bank, 1 billion people don't dispose of (safe) water.
Contaminated water causes 80% of the diseases in the developing world;
4 to 5 million children die from contaminated water.
In the last century the world population increased threefold, the water consumption sevenfold.

The Colorado River and the Huang He don't reach the ocean anymore - they are empty. Access to the sea for nutrients is stopped by dams, which interrupts the food chain in the ocean. Too bad for the fish in the sea; the rivers no longer contain fish anyway. In the San Joaquin Valley (California), the dry emptying of underground water pockets caused a 10-meter collapse of the surface.

Even in deserts there is plenty of groundwater but due to irrigation the reserves are disappearing faster than they can be replenished. Wetlands and oases vanish, reserves get exhausted
In Saudi Arabia the groundwater is being consumed at a level three times the replenishment rate. Libya extracts 3.8 billion m³ of water more from the underground than it gets replaced. In Punjab, India's granary, the water level sinks 7 inches per year. In China the phreatic layer has fallen by 120 feet in the last 4 decades.

China is therefore heading towards a real water crisis: large reservoirs are drying up because of less rainfall, less than average for the last ten years.

This situation is enhanced by the strong economic growth and the water pollution.

Reservoirs are being guarded by soldiers and police!

The average Chinese citizen has about 300 m³ of water per year at his disposal, which is an extreme deficit according to UN criteria. They estimate that 1000 m³ per year is a minimum. Yet China has about 7% of the global amount of water but 20% of the world's population.

We could go on forever with these examples, in many areas of the world

Nowadays, between 30 and 40% of all people suffer from water shortage.
At the same time we keep plundering our reserves in rivers, lakes and in the underground.

The day will come we will have to "repay".

By 2025, 66% of all people will suffer from water shortage or, to put it in another way: there is enough water for only 1/3 of the population, even now.

We just aren't aware yet.

What can we do about it?

The answer is plain, simple and disenchanting: nothing.

We have no way to increase the water supply. This absolutely indispensable product is only available on earth in a limited amount. We have to with what there is. We should only rely on the reserves as a buffer for a temporary shortage. Recycling is once again the magic word.

The most important problem concerning water consists in the fact that there is only a limited offer for an increasing demand due to the population growth on the one hand and the ever-growing consumption of water by the human economical activity.

Since the offer cannot be increased, the balance must be obtained by a decrease in demand.

We have only preventive measures left:

Consume less, pollute less, waste less, keep rivers and lakes clean, spare the phreatic layers, build reservoirs, slow down the discharge of rain by making the rivers meander, instead of turning them into canals, separate rainwater and sewage…

Fortunately the soil in the deserts locally contains sufficient phreatic water in order to enable reforestation.

That water should only be used for that purpose.

8. Global warming

We already mentioned deforestation and desert building as causes for global warming, and we stressed the role of water as moderating element on the climate (see also appendix).

Global warming remains a controversial topic.

What has been proven beyond any doubt? Is there really such a thing as global warming? Is this due to human activity? Are the s. c. green house gasses involved?

Hence: some historical facts and data.

1. In the course of history several periods of global warming have occurred, the interglacials, alternating with cooling periods, the well-known ice-ages (glacials). We may expect the next glacial period in about 15 to 20.000 years.

2. These long-term variations are due to the s.c. Milankovic effect:
    - variations in the eccentricity of the earth orbit (period of 100.000 years);
    - variation of the axis of the earth as compared to its orbital plane, or obliquity (period of 40.000 years);
    - the precession of the earth axis (period of 26.000 years).

    Since these three phenomena have different periods, they interact in a varying way, sometimes reinforcing each other, sometimes counteracting.

3. There are also variations with a shorter period. They are called the cycles of Dansgaard-Oeschger (during the glacial periods) and cycles of Bond (during the interglacials such as the actual Holocene or Anthropocene, as it is called recently).

    Both are rapid climate fluctuations with an interval of 950 to 1900 years. The cause of both these phenomena is yet unknown. Maybe as a consequence of the interference of the three Milankovicz parameters?

    Characteristic is a rather rapid warming (a few decades) followed by a slow cooling down (a few centuries).

4. We now know that the Middle Ages were quite warm, while we experienced a "small ice-age" between 1450 and 1750, as we know from Brueghel's paintings. This could mean that we are now experiencing an upward Bond cycle.

In this context human activity would play no role in the rising earth temperature.

On the other hand there are some striking facts...
Let's look at a few of them more closely

1. Reliable institutes such as the IPCC (International Panel on Climate Change) consider that it is very likely that there is a human interference through deforestation, use of fossil fuels and the exhaust of CO_2 and methane.

2. As a result of the melting of the polar ice, the sea level would rise between 18 and 59 cm in 2100 causing major flooding.

3. When the polar ice melts, the white reflecting layer diminishes; less sunlight will be reflected (less albedo of the earth) and more sunlight will be absorbed; this will lead to more and faster global warming. This is, in fact, happening now.

4. The polar ice has melted to such an extent that navigation around the polar ice has become possible through the north-western (Canadian route) and by the north-eastern, Finnish and Russian route, for the first time in 125,000 years.

5. Projections are that by the summer of 2040 the North Pole will be ice-free. Only recently it was thought that this would not happen before 2100.

6. The steam/water or water/ice mixtures have a buffering effect on the fluctuations of the temperature (see appendix). Disappearance of the polar ice could result in an unlimited, not buffered rise of the temperature. This is probably going on now.

7. We assess a diminishing agricultural productivity in the Middle East and India; we see the effect on the eco-systems, on the climatic areas and on the biodiversity; deserts grow bigger, new deserts are formed.

8. Today, we also notice more extreme weather conditions, severe tropical cyclones at a greater distance from the equator. In 1985 we noted 120 meteorological disasters; there were 500 of them in 2006, causing flooding problems for 248 million people.

9. In the Northern Ice Sea the plankton, the base of the entire food chain, is disappearing rapidly.

10. The disappearance of the coral reef, the base for many other forms of life, is going on now.

11. The glaciers in the Alps (the "water-towers" of Europe) are melting. Up to now the ice buffers the rise of the temperature in the Alps. But what will happen afterwards? How much water will be available in summer in the Rhine, the Rhone, the Po, the Danube? What will be the consequences for navigation, irrigation, and fishery, for the cooling systems of adjacent industry?

12. The ten warmest years ever registered all happened after 1990.

13. The average temperature rises very fast, in fact faster than one would expect from the rising phase of the Bond cycle. The polar ice is melting very rapidly; the Bond cycle usually causes a rise in temperature during a few decades but now things are moving much faster.

14. The forecasts are an average rise In temperature of 3 to 9°C by 2100. In comparison: the average temperature is now only 3 to 6°C higher than in the last ice age, about 10.000 years ago.

    Or otherwise: the mean temperature is now just half a degree C° higher than during the last small ice time (1550-1750) A small change in temperature can obviously cause important climate changes.

15. Not only the ice is melting, so is the permafrost. Areas that had been covered by ice for centuries become weak and even turn into swamps. Steep mountain sides also weaken, slide down into the valley, causing huge avalanches. The Tibetan plateau (+5000 m) gets weaker and the recently built railway and highway sink into the soft soil. They need a solid foundation over more than 1000 km. The same happens in Alaska where the oil pipeline is damaged and also needs a new and solid foundation.

What can we conclude?

There is no doubt: global warming is on its way.
The IPCC states that this is a manmade phenomenon.
There are many arguments to state this theory.
Even if the phenomenon should have natural causes, we must react.
In the same way we must do something about the greenhouse gasses.
Even if they wouldn't be the cause of global warming

What can we do about it?

Part of the solution lies in combating desertification:
- Creation of new forests and green areas all over (refer to "menaces" 1 and 2, above)
- Anticipate to the expected negative effects of global warming. This may cost a lot of money e.g. new dikes

If we conclude that global warming is a result of the emission of greenhouse gasses, we do have a broad area of solutions. This will be treated in the next chapter, menace 9.

9. The Greenhouse gasses

Controversy

Fortunately greenhouse gasses occur naturally. Without these natural greenhouse gasses, the temperature on earth would be some 32° C lower: a medium temperature of -18 C instead of +14 C now.

Some people deduct that these gasses have no correlation with the human industrial activity or that at least the human influence is minimal and negligible.

In 1824 the Frenchman J.B.J. Fourier found out that the atmospheric gasses could make the temperature on earth rise.

In 1896 a Swede Svante Arrhenius published some calculations of temperature changes by the variation of carbon dioxide. A double concentration would lead to a warming up of the earth by 4 to 6 degrees. Arrhenius is considered the inventor of the concept of global warming. He received the Nobel Price for chemistry in 1903.

Today most scientists, headed by the IPCC agree with this theory. They are supported by the signature of 168 governments, among them 15 EU-countries.

This conviction is the basis of the protocol of Kyoto.

Kyoto - 1997

Hundred years after Arrhenius this problem is brought to our attention

Kyoto talks about 7 gasses:
- H_2O or water vapour: responsible for 36-70 % of the effect.
- CO_2 or carbon dioxide, 9-26 %
- CH_4 or methane, 4 à 9 %
- O_3 or Ozone, 3 à 7 %
- SF_6 or sulphur hexafluoride
- N_2O or nitrous oxide, laughing gas
- CFK's, chlorine-fluorine-carbon compounds (several compounds)

The gasses mentioned above are classified according to their *global impact*, i.e. the result of their *concentration* and their *specific impact* on the global warming.

SF_6 and CFK's are not natural gasses, they are man-made.

We will focus on the four most important gasses:
H_2O, CO_2, methane and ozone. In addition: some facts about the CFK's and the ozone gap.

How did the greenhouse effect originate?

- The earth surface is heated by the sun through the penetrable atmosphere
- This surface radiates heat i.e. infrared rays
- These longer wavelengths are stopped and absorbed by certain molecules
- The atmospheric layers are hereby heated
- Part of the warming up is radiated into the space and is "lost"
- Another part stays in the atmosphere or is reflected to the earth surface
- This phenomenon is called the greenhouse effect
- Some molecules are natural, others are anthropogenic
- Some interact with each other, others have a positive feed-back
- All these different interactions explain the often unclear, capricious results
- This could be the explanation for the sometimes conflicting conclusions

Here are some gas concentrations (except water vapour), before industrial revolution, and today:

Gas	Before 1750	Today	Increase	Impact
CO_2	280 ppm	384 ppm	104 ppm	1.46
CH_4	700 ppb	1,745 ppb	1,045 ppb	0.48
N_2O	270 ppb	314 ppb	44 ppb	0.15
CFK-12	0	533 ppt	533 ppt	0.17

**Environment? Atmosphere?
What are we talking about?**

If we represent the Earth as a sphere of 1 meter diameter,
then the atmosphere we're living in, the air layer around the Earth,
would be one millimetre thick

One millimetre!

Water vapour

1. Preliminary remark: we are talking about vapour, i.e. water in its gas stage or H_2Og not about clouds, which are made of water drops and/or ice crystals. But these clouds do play an important role, as we will see.

2. Vapour is present naturally as a result of evaporation of the earth's surface, the vegetation, the rain, the volcanoes etc.; the human input is large though, mainly as a consequence of burning oil (see CO_2). Here, by the way, lies part of the answer to the question: what can mankind do about it?

3. The concentration of vapour in the earth's atmosphere is relatively low: if all the steam would drop down all at once, it would produce a water layer of only 25 mm. The average yearly global rainfall amounts to 1 m, or 40 times more. This proves that the steam is swiftly turned-over in the atmosphere. The average lifespan of vapour lasts only 10 days.

4. Some older statistics only took into account the other gasses, not the water vapour that does play the most important role in the global warming: 36 to 70%.

5. Condensation of vapour into water or ice causes rainfall or snow. This occurs at the edge of warm and cold air masses. The formation of vapour needs sun heat, often over hundreds of km² of the sea; this energy is released back when condensation occurs, often on a restricted area. This enormous amount of energy causes turbulences and convections, the source for devastating cyclones.

6. A relatively small increase in vapour concentration can therefore be responsible for a much higher frequency of cyclones. (4 times more in 20 years) These tropical storms will grow stronger and occur at a greater distance from the equator.

7. Furthermore we witness a positive feedback: vapour increases the temperature of the atmosphere which causes more water to turn into vapour, which in its turn causes more greenhouse- effect, etc. Methane shows a similar pattern. Other greenhouse gasses don't display this self-increasing effect, but they do intensify the evaporation of water and eventually the release of methane.

8. More vapour should create more clouds, which reflect the sunlight back into space, thus slowing down the influx of heat. This apparent logic assumption has not been proven yet.

9. Once again water (as vapour or ice) shows its regulating, buffering effect on climate and environment. Vapour is by all accounts the most important greenhouse gas.

Carbon dioxide

1. Carbonic acid, as it is also named or CO_2 has the second most important impact on the environment of all gasses. We should refer to carbonic acid only for H_2CO_3. CO_2 was discovered by the Fleming Jan Baptist van Helmont (°Brussel 1577 +Vilvoorde 1644). He discovered something new which he called "gas", unknown until then. He named it Sylvester gas. He was named the Leonardo da Vinci from Brussels and he is considered to have discovered the photosynthesis.

2. Carbon dioxide appears abundantly in nature, but the concentration increased: from 280 ppm before the industrial revolution to 383 ppm nowadays. This is the highest concentration since 420,000 years and a further increase between 490 and 1260 ppm is expected by 2100.

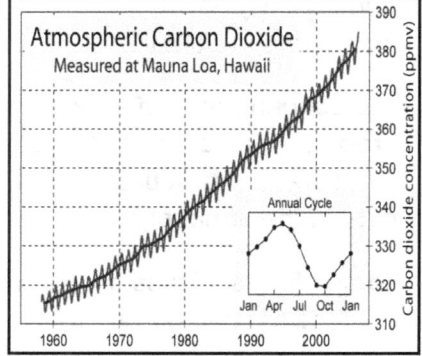

Source: Wikipedia

3. Carbonic dioxide is a very special gas. On the one hand it is a waste product from humans and animals who exhale CO2, on the other hand it is a basic necessity for plants who, thanks to chlorophyll, turn CO2 and water into their texture and produce the oxygen, so badly needed by the same humans and animals. This constitutes the perfect cycle, going on for millions of years.

4. The motor here is the sun, for free. The amount of energy that reaches the earth is 9.000 times larger than what the whole mankind needs. The sun produces directly and indirectly (through wind, water power, biomass etc.) 99,9 % of all renewable energy on earth, thanks to chlorophyll. Chlorophyll is undoubtedly the second most important substance on earth. (notice: fossil fuel is really stored solar energy).

5. For thousands of years this cycle was faultless. But more people live on earth, and they produce more CO_2, especially by using fossil fuel. The industrial emission takes about 6% of the total amount. At the same time forests are being destroyed to make place for the growing human masses, thus diminishing the absorbing capacity. An average family yearly produces 15 tons of carbon, not counting the use of cars, travels etc. Hence, it is not surprising that the CO_2 concentration rises. Vegetation in Europe absorbs about 125 million tons of CO_2 every year. We cannot live without this vegetation.

6. Carbonic acid is also stored in the oceans, where it contributes to acidification of the water and to the destruction of the coral reef that constitutes an important habitat for a whole range of animals and plants.

7. The USA is the biggest polluter: 4% of the world population produce 25% of the total CO_2 emission. Their coal power plants alone produce 250 billion tons of CO^2, the cars some 150 billion tons.

8. The combustion of nonane, one of the hydrocarbons that constitute petrol, goes as follows: $C_9H_{20} + 14\ O_2 = 9\ CO_2 + 10\ H_2O$. In other words: 1 mol of nonane needs 14 mol of oxygen in order to produce 9 mol of carbon dioxide and 10 mol of water. Other hydrocarbons, such as heptane, octane etc. are similar. H_2O and CO_2 are in fact useless waste products; as greenhouse gas they are even a nuisance. Fossil fuels deserve a better use: they constitute the basis for chemical products (see Chapter 6 - Energy)

9. As a comparison: the planet Venus, about the same size as the earth, has an atmosphere of about 96% CO_2 and its surface temperature is 464° C. Another proof of the impact of CO_2 on the greenhouse effect.

10. Paleo-climatic studies over the last 420.000 years show:
    - The concentration of CO_2 was never as high as today
    - The rapid rise of this century (80 ppm) cannot be attributed to the rise of temperature alone
    - The oceans will absorb less CO_2, more will remain in the atmosphere.

11. Hence scientists from the IPCC conclude that the global warming is a fact and that this is due to man made CO_2. (+95% probability) By the way: even if this would not be the case, we should decrease the emission of CO_2 and stop burning valuable hydrocarbons. They can be of better use.

12. What can we do?
    - Promote the absorption of CO_2, e.g. by creating more forests
    - Diminish emissions by using less fuel
    - Use renewable energy only: should be our target ASAP
    - However, this may be hard to attain, if world population keeps growing

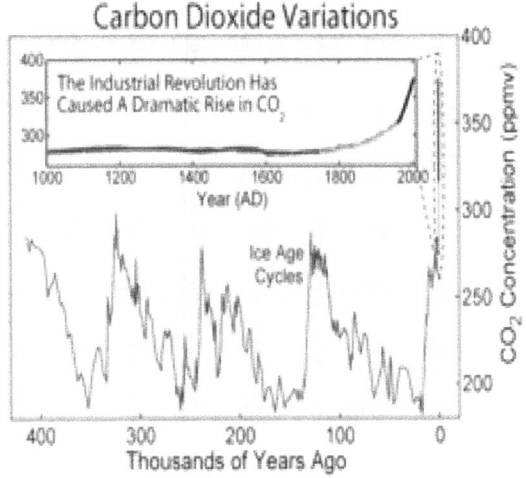

Source: Wikipedia

Methane

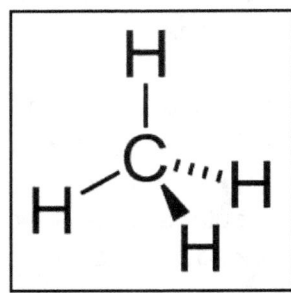

1. Methane, marsh gas, CH_4 is a natural gas. It was discovered by Alessandro Volta in 1778. It is the most simple organic compound, present in wet regions, marshes and bogs. By global warming and thaw the permafrost (in Siberia and Canada) releases more CH_4 than before. Again: the global warming sustains and even invigorates itself: global warming increases the amount of CH_4 and the increase of CH_4 influences the greenhouse effect, etc.

2. Since 1750 the concentration has increased by 150%, half of it manmade. Agriculture and especially farming produce a great quantity of methane in the stomach and intestine of cattle. The disposal of waste and the production or the use of mineral fuels such as natural gas also have an influence on the methane level.

3. Under a certain pressure and temperature conditions methane forms a complex with water molecules: called methane hydrate or clathrate. Eventually this may lead to the release of methane e.g. in the permafrost areas. The melting permafrost offers the possibility to bacteria to turn more organic remains into even more methane. Even in the deepest sea bottom clathrates can be found.

4. Methane is a greenhouse gas, about 20 times as powerful as CO_2. The concentration in the atmosphere however is about 700 ppb (parts per billion) or circa 1/600 of the entire CO_2 - concentration. Its lifetime in the atmosphere is about 10 years, which is less than most toxic gasses.

5. How can we reduce the input of methane? By taking measures in the farming and waste-disposal industry, by using less natural gas. Yet it is unlikely that tomorrow 8 or 10 billion people will succeed in producing (or wasting) less CH_4 than the 6.8 billion today. The gravity of the problem, obviously, is enhanced by the amount of the world population.

CFK's and the Ozone

1. Ozone is an unstable compound of 3 oxygen atoms: O3. It is an extremely powerful oxidant and reacts vehemently with all kinds of materials like rubber, metals etc.
   Ozone was discovered in 1840 by Christian Friedrich Schönbein.

2. It is generated in the atmosphere by electric discharges (lightning), but especially traffic plays a very negative role. The mixture of hydrocarbons (like gasoline) with CO (carbonic monoxide) and nitrogen oxides (NxOx) is called smog. In this smog ozone is produced under the influence of the sunlight and heat. This ozone is very harmful for humans, animals and vegetation.

3. In the stratosphere, ozone is formed by the UV light of the sun. Ozone is very useful here because it stops the UV radiation. Otherwise skin cancer would strike more humans, and the plankton in the oceans would be more damaged. Yet plankton is the basis of the food chain of the sea.

4. This stratospheric ozone layer is being affected by gasses like the CFK's. This explains the ozone gap over the South Pole. These gasses were banned by the protocol of Montreal in 1989, with a positive result today *(CFK's 1 million tons 1988, 10,000 tons in 2008)*.

5. The ozone gap was discovered in 1985, but already in 1974 scientists, after viewing the results of their measurements, thought that there was something wrong with their instruments. Also in 1974 some American scientists discovered that these CFK's played an active role in the destruction of the ozone layer.

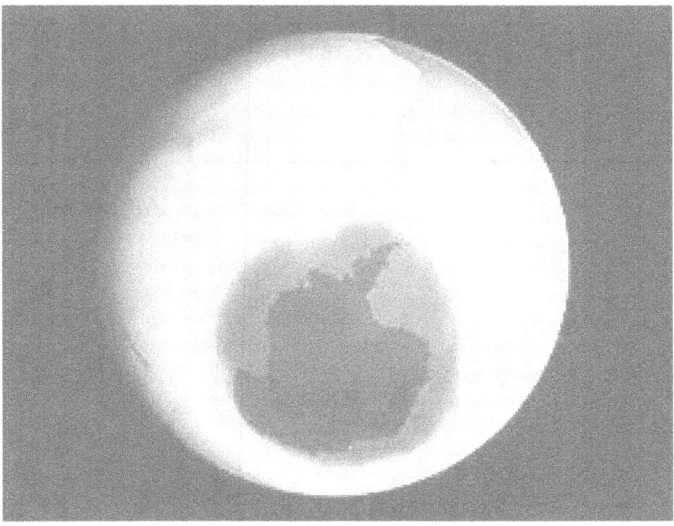

The ozone gap (light gray) above the Antarctic (dark gray)
Source: Wikipedia

Other gasses

1. There are dozens of organic compounds containing chlorine or fluorine, or both Cl and F. They may differ in intensity and duration of action. Also their Global Warming Potential (GWP) may differ. Just to get an idea, take a look at the small list below. Please notice that the GWP of CO_2 was used as standard. Hence, it appears that sulphur hexafluoride has a GWP 23.900 times higher than CO_2.

   SF6 is used as insulation gas in high-tension switches and in the chip industry. In 1995 the emission amounted to an equivalent of 1,5 megatons of CO_2. Its life expectancy in the atmosphere is about 3200 years. The life expectancy of laughing gas is 120 years, its GWP is 310.

2. It is quite obvious: some gasses are more powerful and/or have a longer life expectancy than CO_2. Therefore they must be carefully monitored, their use must be restricted and they must not be released into the atmosphere. The problems with these materials are common knowledge and are well documented. The necessary measures have been taken and are being implemented. First successes have been obtained e.g. in the case of the ozone gap. This should not constitute a major problem in future.

Gas	Chem. formula	Global Warming Potential
Carbon dioxide	CO_2	1
Trifluoromethane	CHF_3	11,700
Pentafluoro-ethane	C_2HF_5	2,800
Laughing gas	N_2O	310
Sulphur hexafluoride	SF_6	23,900

Sources: - IPCC, Klimaat Nederland, Wikipedia.

Note:
Recently, an even stronger greenhouse gas was found: poly-fluoro-tributyl-amine: C12-F27-N (2014)

The nine main challenges: Summary & conclusions

Undoubtedly there are many more problems, but these nine threats certainly are the most important ones. For each of them, three questions arise:
1. The seriousness, i.e. how lethal, how dangerous is the threat?
2. The urgency: how much time is left to us to get into action?
3. Are there causal factors that occur more often?

1. The seriousness

It is not necessary, that all nine the threats make alarm; it is not necessary, that a majority makes alarm (e.g. 5 out of the 9). Eight of them are so *vitally important*, that a signal from any one of them suffices to justify a general alarm, that requires an immediate and appropriate action, because doing nothing would lead to a catastrophe for mankind and environment.

If we classify these threats from A to D
- A - meaning very threatening
- B - threatening
- C - less dangerous
- D - not dangerous

8 of them should be classified as A.
Only the depletion of minerals can be classified as C, or less dangerous

2. The urgency

Another important factor is the urgency.
For some menaces we *are too late already* for action. (1)
For others it *will be too late*, when our actions of today become effective. (2)
For others still, the effect of our actions *may be timely*, if we start now, today. (3)
For some, we may have some time left. (4)

	The urgency
1	We are **too late already**.
2	It **will be too late** before our actions come into effect
3	We must **act immediately** before it is too late
4	We have **some time left**

3. The causal factors

Most of the nine threats interact and reinforce each other. In all nine, the *population factor* is omnipresent, sometimes as the provoking factor, another time as the reinforcing element, often as the only causative factor. Some threats can be re-balanced only by lowering the demand because the supply has reached, or even surpassed its limits. This is especially true for the overfishing which is one of the most dramatic elements in this series.

But also in the cases of food shortage, of exhaustion of the natural reserves of fossil energy, and of the water, the population pressure plays a preponderant role. The overpopulation finally, also plays a determinant role in the desert formation, the deforestation, the global warming and the greenhouse gasses.
On top of that, the great leap in consumption in the third world is still to come: they also want to live and consume like we do in the West and in the USA, with a larger ecological footprint.
This statement confirms and reinforces the impact of population pressure. The population pressure is the main determining element, which should guide our behaviour, if we want to preserve the future of humans.
Animals, birds for instance, can propagate unbridled, until hunger "spontaneously" reduces their number, creating in this way "a natural balance". Is THAT what we wish for mankind???

Population pressure is the tenth menace.

	Seriousness and urgency	
1	Desertification	A / 1-2
2	Deforestation	A / 1-2
3	Food shortages	A / 1-2
4	Overfishing	A / 1
5	Depletion of reserves: ore and minerals	C / 4
6	Depletion of reserves: energy carriers	A / 2
7	Depletion of reserves: water	A / 1-2
8	Global warming	A / 2-3
9	The greenhouse gasses (+ the ozone gap)	A / 2-3
10	Population pressure	A / 1

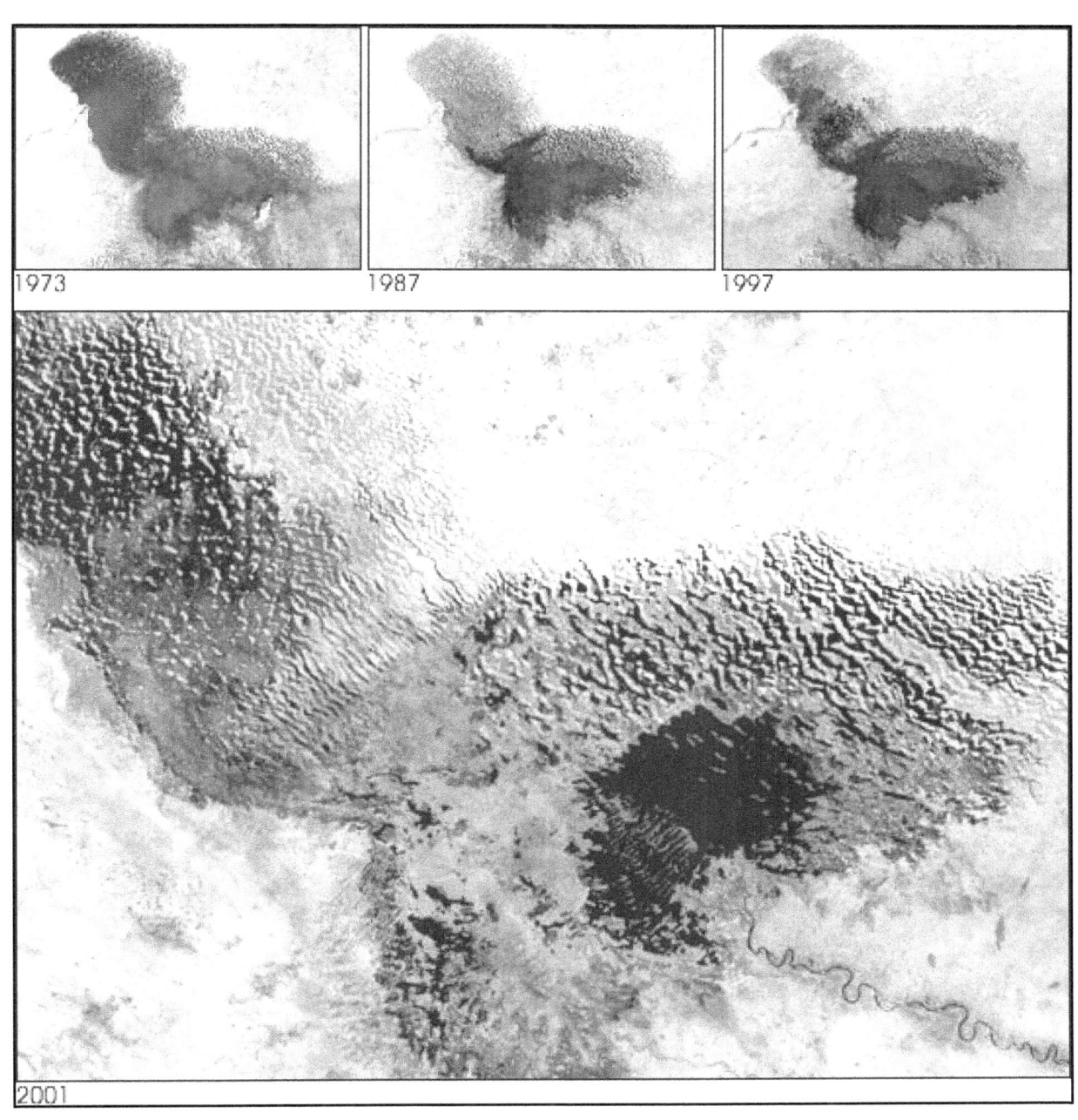

The vanishing Lake Tchad (Africa)
Source: NASA Satellite photos (via Wikipedia)

Chapter 3

The Tenth Menace: Overpopulation

1. Historically
2. The figures today
3. The geographic spread
4. Urbanization
5. Influence of the development level.
6. How to measure the population pressure
7. What is an ecological footprint?
8. What is a population index?
9. How to implement Kyoto

Population pressure

Can we quantify this?

1. Historically

For centuries the world population remained quite stable:
- about 1 billion in 1800
- a slight increase to 1.6 billion in 1900
- then started a spectacular derailment, that goes on today

Source: Rotary RFPD fund

2. Today's figures

The graph shows the world population in 2005 (6.5 billion); but in 2014 we are 7.2 billion.
- For real time statistics, go to http//www.worldometers.info
- Each year the world population increases by some 80 million, equal to the population of Germany or Mexico
- This means 3 per second or 220,000 a day

3. The geographical spread

Out of every 100 world citizens:	60 are Asians
14 Africans
12 Europeans
9 South Americans
5 North Americans

4. Urbanization

Since 2007, more people live in the cities than on the countryside. Every day 180,000 humans move from the countryside to the cities. One out of three townsmen (i.e. one billion) lives in slums in developing countries of Asia or Africa. And their number increases rapidly.

In 1950, there was just one town with more than 10,000,000 inhabitants: New York.

A megalopolis counts more than 10,000,000 inhabitants, OR more than 2,000 persons per km² Nowadays there is one megacity with 35,000,000 inhabitants: Tokyo. In 10 year time, there will be nine megacities with over 20,000,000 inhabitants: Bombay, New Delhi, Mexico, Sao Paolo, New York, Dhaka, Djakarta, Lagos.

In 1950 there were 83 towns with over 1,000,000 inhabitants, now almost 500. In Ethiopia, Malawi en Uganda 90 % of the townsmen live in slums. In 2030, there will be 8.1 billion of humans; 5 billion of them in cities, and 2 billion of them in slums. Of 100 new-borns worldwide today at least 70 will later live in a town, and 20-25 % will be a slum resident. Or in the best case will he become an emigrant.

Nigeria expects 25,000,000 inhabitants in its capital (Lagos) before 2015; 80% of the townsmen in Nigeria live in slums; i.e. 42,000,000 people. China expects that 800 million will live in towns in 2015. In India today almost 160 million humans live in slums.

The F.A.O. warns for serious food shortages *in the slums* of these mega towns. There is a great risk for social disorganization, political conflicts, criminality and natural disasters. Slums have no property rights, no urbanisation nor planning; there is no police, no fire department, although the slum is not built with fireproof material! There is no school, no sanitary infrastructure,...

And most of all: chances are very small, that someone can escape this situation, and get a better education and qualification.

The hopeless will revolt, if we do not intervene.
Meanwhile, the tendency to emigrate remains very high.
Slum = bidonville = favela = despair. Emigration seems the only solution.

5. The level of development

	Population		
	1950	2030	2050
Industrialised countries	0.8 billion	1 billion	1.1 billion
Developing countries	1.9 billion	7 billion	8.0 billion
In slums	0	**2 billion**	**3 billion?**

The data are obvious: in comparison with 1950, the rich industrialised countries remain *almost status quo* at 1 billion. The poor, developing countries rise from 1.9 (1950) to 8 billion in 2050.

It seems that the richer countries have less children and get richer, while the poor countries have many children and (thus) stay poor. And live in slums. And even get poorer.

An increase of 1 % of the BNP of a developing country may look quite an achievement, but how can this effect the standard of living, if meanwhile the population rose by 1,8 % ?

This table shows the number of children per woman in 10 countries (out of 195 countries). The numbers speak for themselves. In order to have a stable population, about 2.1 births per woman are needed. The table illustrates clearly the discrepancy between the first five, and the last five countries. The last five may not be representative, because they are city-states, or states who recently had severe political or social problems.

Children per Woman, 2008		
1	Mali	7.34
2	Niger	7.29
3	Uganda	6.81
4	Somalia	6.60
5	Afghanistan	6.58
...		
191	Ukraine	1.22
192	South Korea	1.21
193	Belarus	1.20
194	Hong-Kong	0.97
195	Macao	0.91

Source: U.N. Population Prospects

6. How do we measure population pressure?

6.1. Historic grounds

On the Rotary RFPD fund World population growth graph, a few pages ago, one sees that the derailment began about 1920. It is clear that the vertical growth of population (x3 in 70-80 years) is not sustainable. In many parts of the Earth, the population pressure has reached unbearable proportions.

We can no longer ignore, that **overpopulation IS a problem:** more and more humans on a limited surface, with a limited capacity of space, of water, arable soil, forest, nature, etc.
And the process goes on.

For a first, rough estimate of what a "normal" population pressure could be, we can go back to the figures of 1920, before the derailment originated.

This means about 2.3 billion people, about 2/3 less, about 67 % less than today.
Historic arguments, of course, are insufficient.
But there are many more arguments, which do have a rational value.

6.2. Desert, reforestation

In those paragraphs we saw that lots of forests are needed, for nature as well as for forestry; that pre desert areas like the Sahel should be evacuated, that the deforestation in the Tropics – and elsewhere – should be stopped, and even reversed.

We saw, that *the offer* of space cannot grow; that on the contrary it should be reduced; that hence *the demand* is the only variable we can control. And that means: the population pressure.

We will read more about these items in the paragraph "ecological footprint".

6.3. Food shortages

Approximately 1 billion people suffer from malnutrition; 2.8 billion people live below poverty level. Furthermore, we must realize that parts of the actual arable surface will have to be converted into forest, in order to enhance the CO_2-O_2 equilibrium, the water household and the wildlife. Or to serve as a basis for alternative energy, windmills, solar panels, bio-energy, and the like.
In other words: the arable surface will shrink, not grow. On the other hand, the production of crops per acre has reached its limits already; so the supply of food is doomed to diminish now or in the near future.

The shrinking offer will necessitate a shrinking demand, by a shrinking population pressure. There is no alternative.

6.4. Overfishing

Here, the situation is frankly dramatic; the fishing grounds have been looted in a way that a moratorium is urgently needed, for years to come, in order to give the brood stocks a hypothetical chance to recover. Let there be no doubt about it.

In other words: the offer is extremely low, and a recovery cannot be expected before several years or even decennia.

Hence, it is clear that a major part of mankind will have to do without fish as the basis of nutrition. And for some people, there will be no substitute, no alternative. We can therefore expect serious problems: famine, starvation, food riots. Even the better-off people will have to accept a limitation of the fish supply, without lamenting, and accepting higher prices, until the brood allows a higher consumption again.

All alternatives lead to a disaster.

When we take overfishing as the standard for the overpopulation, the score would be 80 – 85 – 90 %. And that is indeed the score of overpopulation, if we do not succeed, to let the fish stocks recover. Optimism is therefore only justified in many years, or decennia. If ever…

But, even in the best case, when fish is present in the oceans again in quantities like 50-60 years ago, we are still not secure: obviously the stocks of 60 years ago did not suffice to allow the massive fishing. And in those days, there were 3 billion humans less to feed than today.
In other words: even in the case of full recovery of the fish stocks, the population pressure will be at least 60 – 70 % higher than bearable.

6.5. Reserves: ores and minerals

The products produced in this activity branch, mostly metals, can easily be recycled, even several times.
Therefore, they do not cause important ecological problems. Still, overpopulation may necessitate new mines to be opened. In this way, population growth does play a (minor) ecological role.

6.6. Reserves: energy carriers

In order to survive in the long-run, we should exclusively rely on renewable energy, because either the fossil reserves get exhausted (a matter of only a few decades), or when we have sooner discovered, that fossil fuel has to be reserved for other, more "noble" purposes.
In other words: we must replace *90 % of our energy consumption* of today, originating from fossil fuel, by energy coming from the sun, the wind, the tide, eventually also from bio-diesel and bio ethanol.

Quite a task!

To capture this energy we will need a large infrastructure and more space still; and space is, as we know, an important aspect in population pressure. We will deal with this more extensive, in the paragraph "ecological footprint"

6.7. Reserves: water

6.8. Global warming

6.9. Greenhouse gasses

These three items were treated in former paragraphs; they will be spoken about also in next paragraph: the ecological footprint.

7. What is an ecological footprint?

The concept "ecological footprint" is based on the fact that all renewable sources of mankind originate from the Earth.

The ecological footprint means: the surface of land and water an average local human needs, in order to meet his needs for materials, energy, and processing of his waste.

It is quite possible, that part of man's footprint is in another country; or even in an other continent. The footprint is measured in hectares of biological active space, with an output equal to the world average.

In that sense our biosphere measures about **11.2 billion hectares of biological active space** or about ¼ of the total surface of the earth. That 's all there is: 11.2 billion hectares.
There are 2.3 billion hectares of water (seas and lakes) and **8.8 billion hectares of solid land**.

The solid land = **8.6 billion hectares**.
- 1.5 billion hectares of farming land
- 3.5 billion hectares of grass land
- 3.6 billion hectares of forests
- + 0.2 billion hectares of built up land (houses, roads, industrial plants,...)

These 11.2 billion hectares for a world population means (in 2007) a space of 1,8 hectares available per person. But in 2007, the footprint was 2.7 hectares. On average!

This makes clear that we are living in "overshoot", and it also shows the importance and the magnitude of the overshoot. We must also not forget that our data are rather optimistic.

In reality, things are even worse.

And this was 2007 - at the beginning of 2014, things are much worse again.
The world cannot go on living like this.

The footprint varies enormously from country to country:
- 10 hectares per inhabitant in the UAE, USA or Kuwait
- less than 1 hectare in Haiti, Somalia, Afghanistan.

If we compare the footprint with the available bio-productive surface of a country, we can determine which country shows a deficit. With the countries mentioned above, there are also Japan and the UK. They take advantage of the lower consumption in other countries.

**The ecological footprint is not a precision instrument.
It underestimates reality. But as yet there is no better.**

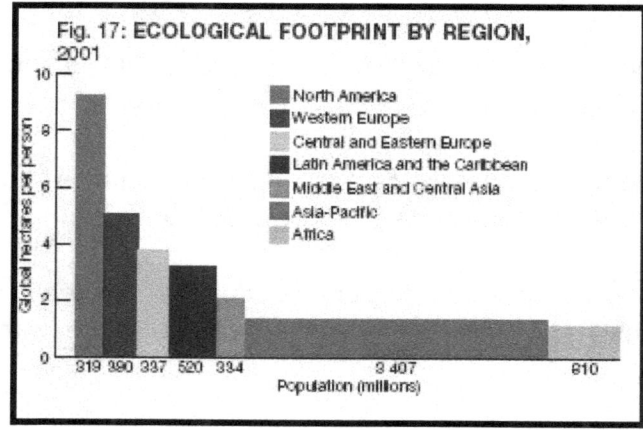

Source: footprintnetwork

If all world citizens would have the same pattern of consumption and waste production as the Americans, we would need FIVE planets Earth; and if everybody would behave like the Europeans, we would need THREE planets.

We all know there are no 3 or 5 planets; we only have one unique planet Earth.
Whether we like to hear it or not: the only solution lies on the demand side. This means: consume less by lessening the population pressure. The only alternative would be to lower our consumption pattern to the level of the third world of today.

Overshoot Day

This is the date on which the reserves for the whole current year are already depleted, consumed.

The first Overshoot Day was 31 December 1986.
In 1996 mankind consumed 15 % more than Earth could offer; the Overshoot Day was in November back then.

In 2014, the Overshoot Day was August 19. On that day the total supply and reserves for the year were used, consumed; the remaining days of 2014 we were living "on credit", borrowing from the future.

(Source: Global Footprint Network)

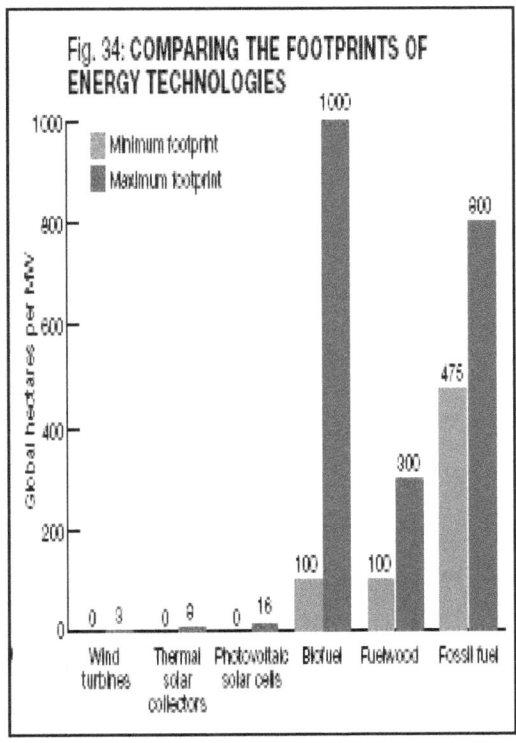

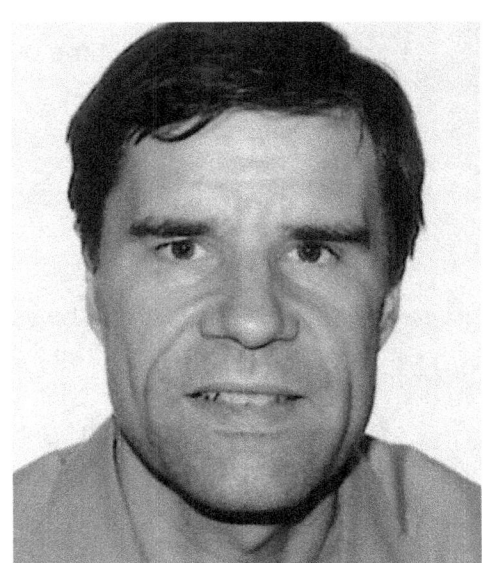

Dr. Mathis Wackernagel,
Executive Director
of Global Footprint Network

This graph shows the footprint of some energy sources. Note the poor performance of fossil fuel, and the even poorer score of bio-fuel.

It is obvious that bio-fuel is not the right answer for the problems of Earth.

8. What is an ecological population index (EPI)?

The ecological population index is the ratio between local footprint and available surface. This EPI can be calculated for each country, for every region, and for the Earth as a whole.
This E.P-index must allow us to judge the actual situation more precisely.

Until recently we had only rough estimates at our disposal. Some were already spoken of in the former pages:

- In 1985 a report by the Brown university, a well respected university of Providence, R.I., USA, pointed out that the earth could only afford 2,8 billion people living the American way of life. This statement is very relevant in view of the fact that the poor countries attempt to consume "the American way", and this also shows that the world population is about 60% higher than what can be "tolerated".

- Other statistics show that mankind consumes and pollutes three times more than what the earth can digest. In other words there should be three planets Earth, or put otherwise: the population pressure is 67% too high.

- If we take in account the overshoot day, our consumption is 33% higher than what can be regenerated.

- If we take the overfishing as the criterion, then the magnitude of overpopulation must be valued temporarily at 80-90 %, and after the hypothetical recovery of the fish stocks at 60-70 % still.

- The problem "water" shows that about the 2/3, or 67 % of humans today suffer from shortage (1/3), or actually consume the reserves (also about 1/3). The reserves will have to be replaced.
The offer of water can't grow, so we can only tackle the demand.

- Other estimates, based on food shortage, greenhouse gasses, energy, lead to the same figures: an overpopulation of about 60-70 %. If we take into account, that fossil fuel will have to be replaced completely; these estimates may be too optimistic.

But: these were just guesses, estimates. Maybe the EPI can be more precise, being the ratio between the footprint and the available surface. The surface can easily be determined, but how do we rate the footprint?

- An old statistic from FAO assesses the foot print at about 3500 m², based on the surface needed to catch enough solar energy in order to produce enough food.

- Yet man needs surface not only for agriculture, but also for his entire complex society.

And this surface varies largely according to his level of development.

- The footprint is:
    - For an African: 1.2 hectares = 12,000 m²
    - For a European: 5.1 hectares = 51,000 m²
    - For an American: 10 hectares = 100,000 m²
    - The average: 3.5 hectares = 35,000 m², an optimistic guess.

Chances are that the world takes over the European way of life, that the Americans lower their consumption by 49 %; but the European footprint is 51,000 m², quite a bit more than the 35,000 m² of the optimistic guess. We will take both figures (35,000 and 51,000) as basis for further calculation.

- The global available surface is, as we saw before, 11.2 billion hectares (2.3 billion hectares water and 8.8 billion hectares land); without the built-up-land respectively 11.0 and 8.6 billion hectares.
  So these are our figures: Biosphere 11.2 billion hectares
  World population: 6.8 billion (2009)
  Foot print: 3.5 hectares and 5.1 hectares.

- Please note, that these figures are quite moderate:
    - The biosphere consists of land – and water surfaces;
    - The world population, before our measures produce any effect, will be much higher, than the 6.8 bln. taken into account;
    - The foot prints taken into account, are rather optimistic.

- The formula of the **EP-index** is:

$$\frac{\text{population} \times \text{footprint}}{\text{surface}}$$

Courtesy of Prof.Em.Louis De Backer.
Univ. Louvain-la-Neuve, Belgium.

There are two possibilities:

$$\frac{6.8 \text{ billion} \times 3.5 \text{ hectares}}{11 \text{ billion hectares}}$$

EP-index == 2.160 **rounded 2.2**

$$\frac{6.8 \text{ billion} \times 5.1 \text{ hectares}}{11 \text{ billion hectares}}$$

EB-index == 3.468 **rounded 3.5**

- That means, that at the lowest estimate, we need 2.2 planets to sustain our way of living; in other words: population pressure is 55 % too high.

- This option still implies, that Americans reduce their consumption and wasting by 60% and the Europeans by 30 %, and that the billions of "others" will not consume more, than the USA and EU will consume then. And that the world population hasn't grown by then. That is a lot of conditions!

- The second hypothesis is more plausible: Americans adopt the European way of life, and the developing countries do the same. But in the meanwhile the world population must not have risen.

- Even under these severe conditions, we will need 3.5 planets for our footprint.

- In other words, as we only have one unique Earth: population pressure is about 70 % too high.

9. If we implement Kyoto

The Kyoto protocol came too late (1997), and it was too soft. The original plan aimed at a reduction of the emissions at 5 % below the 1999 level before 2008-2012
The plan was counteracted by some countries (mainly the USA), it foresaw too many exceptions, e.g. for the developing countries; it only came into effect on 16 February 2005, when Russia, as the 55-th nation, signed in October 2004.

The Kyoto protocol was modified in 2007. It now proposes an emission reduction of 20% in 20 years. Again: not radical enough, knowing that in 20 years the world population will have grown by 1.6 billion or 23 %, and that their individual footprint, meanwhile, will at least have been multiplied. Otherwise: world consumption and emission will double or triple in 20 years. It will be a very hard task, to force a reduction of 20 %.

A plan, that rightfully allows 80 % of the world (the developing countries) to increase their standard of living, and to increase their consumption and emissions, must, as a counter part, demand very severe reductions from the other 20 % of world citizens.

But a reduction of 20% by 20% of the citizens will not produce great effect. These 20% citizens would have to cut their individual and collective consumption more severely.
Will they accept this??

Let us go deeper into this aspect of the Kyoto protocol

1. We have seen, that today 90 % of our energy comes from fossil reserves, who will be extinct in a few decennia. We will have to look for other sources, renewable sources. This would bring about a reduction of CO_2 and other gasses by 80-90 %.
   We can, and must, live with an emission 80-90 % lower, not 20%, as demanded by Kyoto. For the sector "energy" this is possible; it is even mandatory.

2. It is not just about energy.
   There is the overfishing, the consumption of water, the occupation of the soil, the desertification, and the deforestation. These items are hardly mentioned in Kyoto 2007. Yet all these items must be thoroughly be taken care of: fish, water, food, forestry,...
   The required correction will be 50 % or more, for some even 70 of 80 %.

3. In other words: if Kyoto demands a reduction of 80-90 % of the energy carriers, a "greater Kyoto", that tackles the 9 menaces, should go for an all over reduction of 70 - 80 %. For some items this may look exaggerated, but we must realise, that some of them are so important, that they impose their heavy score to all 9 of them. An over all reduction of 70-80 % therefore seems unavoidable: consumption, emission, etc...

4. Once again, we must take into account the "natural" growth of world population. This "natural" growth means 1.6 billion people in 20 years. Twenty years is a common perspective, a common prospect for planning. If we do not act now, the situation in 20 years will be much more difficult.

5. A reduction of emissions of such an extent is hardly bearable; for every individual it would mean a return to the consumption pattern of before the industrial revolution. We can assume that nobody will be prepared to accept this sacrifice willingly.
Acceptance and cooperation will be very poor.
And so will be the feasibility.

6. But there are alternatives.
If we diminish the individual consumption, and at the same time diminish the global number of interested parties, then the individual effort becomes bearable: everybody has to cut less. This prospect may look acceptable. The population numbers have to change reversely proportional to the willingness of the individual to cut his consumption and pollution. That is our core argument.

7. If we all have to cut 70%, whilst population stays unaltered, then everybody has to cut 70 % individually. That will not be accepted.

8. If, on the other hand, the number of interested parties were halved, then everybody's effort could be halved as well. We would still obtain an overall reduction of 75 % of the global footprint, which was our primary aim. If the population pressure would decrease by 75 %, every individual could keep his consumption level, and we would still achieve our goal: – 75 % emission.
Ceteris paribus.

9. As we can expect that each individual will not willingly reduce his footprint by 75 %, we will have to consider the scenario of an equivalent reduction in population.
There is no other way.

10. This should not come as a surprise, since we formulated the same conclusions about all nine of the threats.

Final conclusion:

In all nine menaces the overpopulation plays a predominant role.

Several approaches all display a similar result.

These are astonishing numbers.
Are they mathematically exact?
Probably not - but they don't have to be.

Even if the figures have an error of 50 % towards mathematical accuracy, conclusions would still be the same.

This overpopulation pressure must be evaluated at about 70 %.
(and even more, for some important items)

THE FUNDAMENTAL CAUSE OF OUR PROBLEMS IS POPULATION PRESSURE.

That is what we have to deal with.

Starting now.

But how?

Chapter 4

Population pressure
How to deal with it?

1. Four angles of incidence
2. Emigration vs. immigration
3. Birth rate vs. mortality

1. Four angles of incidence

- The evolution of a population is conditioned by 4 variables, which two-by-two are each other's counterpart: birth rate vs. mortality and immigration vs. emigration.

- This is a very sensitive matter, since it refers to moral, religious and ethical principles. It also has connections to family law, freedom of the individual, solidarity and humanity.

- It is therefore imperative to judge and apply our findings with the highest possible standards, as should be expected from a civilised and ethical society. For the benefit of all and taking into account the sensibility of all. But still without jeopardizing the efficiency.

2. Emigration vs. immigration.

2.1. *Differentiating between immigrants and fugitives*

- Fugitives are people who temporarily leave their homeland because of a war, or analogous recognised reasons; without any doubt they must, for humanitarian reasons, be admitted, even without legal "papers". They come from an area that is a recognised war zone; they must be lodged, they must receive financial support, or a temporary job, in order to be self-sufficient. A temporary permit of residence, eventually an identity card, renewable every 6 months, can help. After a few years (five?), the fugitive may ask naturalization, or – better still – return to his homeland, that may be safe again.

- An immigrant (or applicant) demands quite a different approach. He is not seeking asylum, he is no fugitive, he is an economic migrant, who is not coming temporarily, but who wishes to stay permanently in the guest land. He may be expected to show an identity card, and a visa issued by a European embassy in his homeland.
  This would ensure a humane reception in the guest-land. His arrival would be foreseen, social measures could have been put in place, housing provided, etc.

  An immigrant who presents himself at the border-control without the two documents (ID-card and visa) will not be admitted. He does not leave the international zone. These rules greatly facilitate the job of border-control. But there would be no internment, no locking-up in a (temporary) asylum, no yearlong legal procedure, no hunger strikes, no misery, no stress.

  On the other hand: if accepted, the immigrant will integrate in the culture of the guest land. A colonist is an immigrant who does not adapt, or does not want to adapt to the culture of the guest land.

2.2. The role of emigration/immigration

- Emigration was widespread during colonialism. Nowadays it is hardly a method to tackle overpopulation. It is hardly acceptable, and therefore very little useful as instrument, because the other countries also suffer from overpopulation.

- The same goes for immigration. One can't expect that a country, which decided to control its population, still limitlessly allows immigrants from countries that do not (wish to) limit their own? A limited selective (economic) immigration can at all times be accepted, e.g. with the European Blue Card.

- In other terms: migration brings no global solution. At the most, it can bring about some local improvement. In view of global solidarity, it would be unfair and improper to count on migration in order to find a solution for global overpopulation.

- On the other hand, measures about migration may meet less (moral) resistance than interventions concerning family planning. Repatriation of migrants, who have run through the legal procedure, and whose demands have been rejected, should be accelerated.

- For a short period, measures about migration can have a result more quickly, than measures about population control.

3. Birth rate vs. mortality

3.1. Mortality

Clearly, it would be unthinkable, unbearable, and completely immoral, to even consider influencing the number of people this way.

But unthinkable does not mean "impossible".

Some even say it is happening now, every day, under our eyes. They point with their finger at Eastern Congo, Darfur, Somalia, and others. And indeed: the riots do look like genocide. We are not just witnessing lootings, plundering, extortions and rapes. The goal seems to be the displacement of the original population, to destroy their habitat and take their place. In "the good old days" a country and its people were occupied, now they must leave or die.

Looking at it closely, there is a lot of truth in these allegations. For conflicts like these, it seems that two ingredients suffice:
- Two ethnical groups
- Overpopulation (East Congo) or relative overpopulation in an arid area: e.g. Sahel

It is clear, that such conflicts will repeat themselves, and will more and more display the character of overt genocide.

Murders on large scale are not a coincidence. That makes it the more stringent to do something about overpopulation, or about the perception of overpopulation,

That is: via an efficient policy of birth control.

3.2. Birth Control

The starting point is, as we by now know:
- measures about migration produce little effect
- measures about genocide are completely unacceptable
- salvation must come from birth control

Another statement: the situation is so serious, that we must find solutions that have two major qualities: they must be completely efficient.
And they must guarantee the highest possible moral and ethical standards.
With respect for the individual freedom, where it is possible; and with concern of the common interest, where it is necessary. Even if compulsory measures are needed.
These so-called "compulsory measures", by the way, were imposed by the developed countries on themselves, unforced when they spontaneously decided to practice birth control.

Third conclusion: in order to keep a population at the same level, 2.1 children per woman are necessary. But in developing countries, there are up to a mean of 7.34 births per woman (Mali). In order to make a population shrink, women must have less than 2 babies. .

Clearly: in those countries the family structure, the birth rate and the sexual habits have to be altered considerably.

Fourth: the spontaneous birth control in the West was possible only after a long social and intellectual development. And even then it took about 60 years to realise.

Chances are, that we can't afford to wait another 60 years, before these countries have reached a development that would allow them to decide by themselves, without external pressure.
Just remember the figures: every year – precisely in these poor counties – there are 80 million humans extra. That means 4.8 billion of humans in 60 years.

Individual freedom, and ancestral habits, will have to give way to the demands by the general interest, which is, by the way, also their own interest.

Which interventions, what levers do we have at our disposal?

That is a theme for next chapter.

Chapter 5

Our levers

The arsenal is limited

*Facing these gigantic worldwide threats
We do not dispose of gigantic means of defence.*

*We will have to do with some "every-day" means,
applied with patience, perseverance and conviction.*

This is all there is

1. *Social and fiscal measures.*
2. *Education of the local population.*
3. *How can we reach them?*
4. *Can we apply "soft coercion"?*

1. Social and fiscal measures

Many **developed countries** offer fiscal allowances for families with children under the form of tax reductions. This is perfectly defensible for one-child-families. If we want to achieve population reduction, i.e. temporarily one child per woman, it does not make sense to continue these subsidies. On top of that, many countries offer birth allowances. Clearly these measures are senseless when the aim is a reduction of population.

Developing countries on the other hand do not dispose of these public aids, but that does not stop them from growing much faster than the rich countries. In other words: giving subsidies does not increase population; not giving them does not hinder population explosion.

Evidently, these allowances and subsidies are fully useless. It would be appropriate, to stop up these allowances, except maybe for the first child; but respecting the actual situation of the actual big families, of course. It would be wise, to foresee a latent period of 9 months or one year after divulgation of any new regulation. Just to make sure...

Anyhow: social and fiscal measures do not have a great impact in developing countries; and they, precisely, are the target group.

2. Education of the local population

Many well-meaning organizations, both official and private, have tried for years, to lower the birth rate in the third world, by counselling and advice. Often these organizations are handicapped by the fact, that they can't dispense contraceptives.

Only advice then; and this advice has to take into account the free choice of the concerned, the local morale, and the local family customs, etc. Certainly no obligation!

In view of the poor social and intellectual development, it can't come as a surprise, that all these well-meant actions have no effect: birth rates remain too high, and the children will continue to live in slums, or find solace in emigration.

The world map on page 68 shows the countries where contraception is current (light grey) and those where it is not (dark). Clearly, a rapid improvement must not be expected, and in the mean time the world problem of overpopulation gets worse every day.

Knowing that population stays stable with 2.1 children per woman, and if we want to go for a drastic reduction, the example of China comes in the picture, where 1 child per woman is the (legal) standard. (in fact there are 1.8 children per woman). The Chinese are forerunners; they should be an example for all.

Of course, this measure (one child per woman) is temporary. It can be cancelled in 30-40 years, or whenever the obtained result seems satisfactory.

This measure will undoubtedly influence the family structure, and lead to more collective education (schools, local councils, kindergarten) in order to avoid spoilt single kids; it will change the conception of the houses, etc. But all this may have a very positive influence.

And, by the way, there is no alternative.

In China, the birth rate remains high: officially 1.8; but presumably there are a lot of non-official births, not registered, of children who officially do not exist. Who can't go to school, can't work, etc. These Pariahs have no future what so ever.

This underlines the necessity of a 100 % secure contraception.

Contraception must be enforced in the same way, worldwide. In the rich West and East, as well as in the poor South.
Everywhere in solidarity.

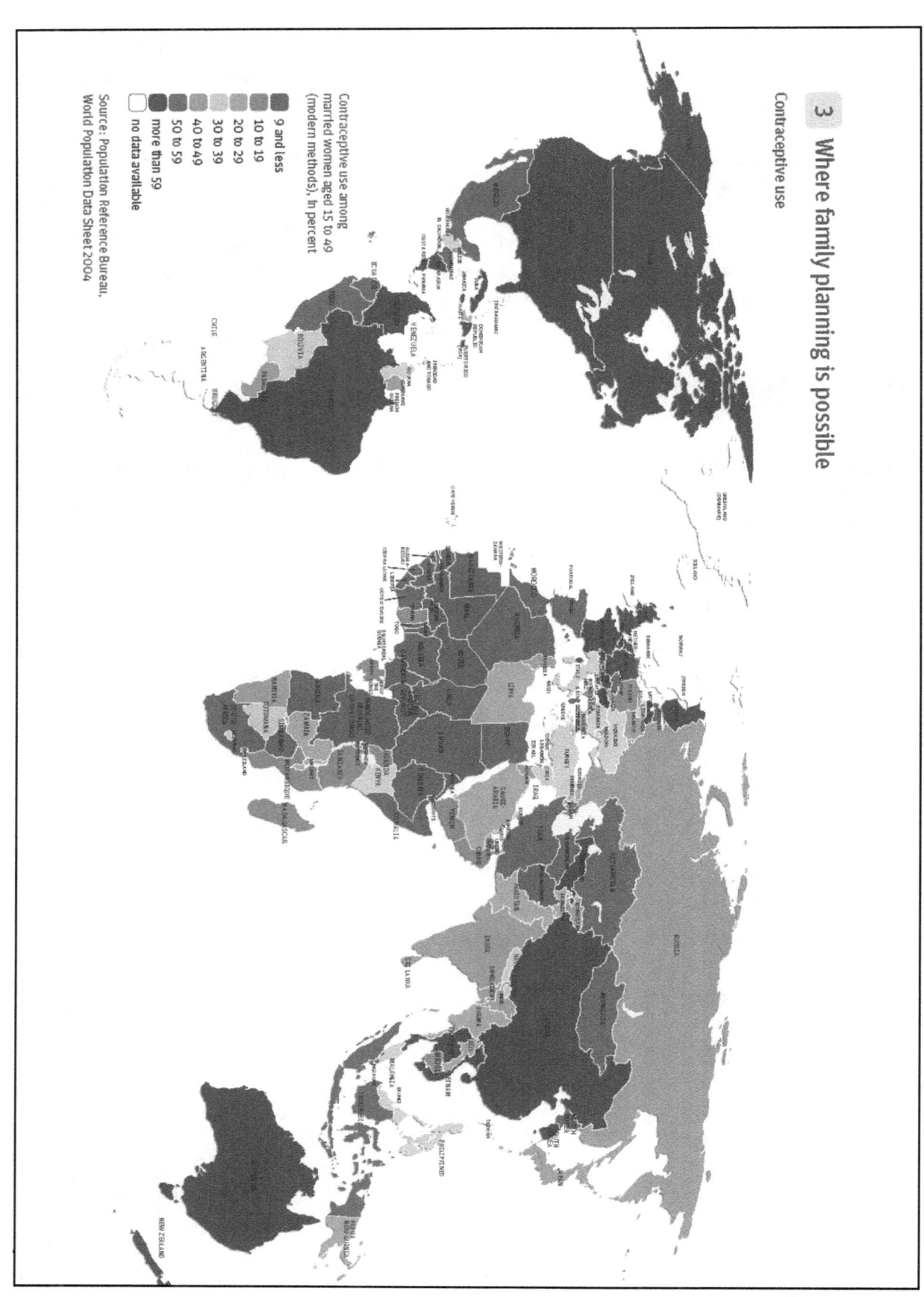

3 Where family planning is possible

Contraceptive use

If the measures that we propose indeed serve general interest only; if the principal beneficiaries precisely are the developing countries, even if they do not as yet understand the importance of the problem; if we distribute the necessary efforts in an equitable way over rich and poor, North and South, then it does seem acceptable, that a certain amount of soft coercion is used, in order to assure the necessary cooperation.

The developing countries can't expect the richer ones to help them with their problems, such as hunger, water shortage, polio, desertification, etc., if they, the poorer countries, are not willing to contribute to the sole problem they *CAN* cope with, i.e. overpopulation, in fact the cause of all other problems of the poor world.

There are other arguments: a developing country can't normally expect to let its citizens emigrate to countries which did take measures to reduce population growth. It is not thinkable that countries that actively practice population control, open their doors to immigrants from countries which do not.

Certainly, there are still more arguments.

Certainly, it is not the right idea, to cut all aid. Aid will remain necessary, for many years to come, from rich to poor, from North to South. That aid must go on. But a certain cooperation may be expected, or demanded, in order to tackle in solidarity the world problem number one: overpopulation; a problem that is the cause of all other problems, who precisely are most acute in these poor, developing countries.

Something else now: countries, that do not implement the Kyoto regulations, could be confronted with higher import taxes, etc.

Yes, there are ways to exert pressure to serve the good cause.

It is not our problem, to discuss this *in extenso*.
UNO and other organisations are more suitable.

In fact: what *can* we do?

1. About energy carriers: switch to renewable energy as soon as possible: biomass, wind, sun et cetera.

2. Fossil sources must serve other, more valuable, purposes.

3. Adopt another way of living that does care for the next generations.

4. Start reforestation (natural forests, and forestry) in all climate zones.

5. Combat desertification, if necessary: resettle population.

6. For ALL problems: mandatory population reduction:

    - Reduction to the figures of 1900-1920 (i.e. minus 70 %)

    - Via emigration/immigration: not very efficient

    - Via a temporary measure of 1 child per mother

    - Via greater tolerance later; but still with birth control

    - i.e. Maximum 2,1 child per woman

To our regret: there is no other way!

Appendix

In all previous chapters the vital importance of water was emphasised.
Water certainly is the most important substance on Earth
- chlorophyll possibly being second.

In this appendix, we will provide you with more facts, data and formulas about water.

Over 100 figures show the irreplaceable role of water
in the homeostasis of the atmosphere we live in,
in the origin of the weather and of the seasons, in precipitation and evaporation,
in the histology of plants, animals, humans; who contain up to 60-85 % water.

Furthermore, there is no substitute for water,
and water is available in limited amounts only.
WATER must therefore remain a subject of our special attention and interest.

The appendix reflects this interest.
It is a speech, given in the Rotary Club Dendermonde
on January 4, 1997.
The speech pleads for a thorough debate about overpopulation.
This plea was not heard then; it is not now (November 2008)

Dawn is over

Some facts about
WATER

Of all substances necessary for life, water is without any doubt the most important. One might think about oxygen, but many creatures are anaerobic, they do not need oxygen; one might think about chlorophyll but many creatures do not need chlorophyll.

But no creature can do without water.
Life originated in water, the first human settlements around water supplies.

I would like to provide you with some facts and figures, some data and also some reflections, that may lead to conclusions, to a sort of diagnosis, to the identification of problems, connected with this essential substance, and by doing so, might lead to formulating the possible solutions. Outside of the realm of biology also, water is of utmost importance. This is due to its massive presence on Earth, and to its absolutely unique properties. We will go deeper into this topic later.

The use of water

Water has many uses:
- Solvent, catalyser
- Standard for several physical unities: the litre, the calorie
- Standard for comparison of physical properties, specific weight, viscosity
- Medium for transport of materials, for the removal of waste products
- Dilution and cooling agent
- Reagent: e.g. the production (cracking) of petroleum: 1 litre takes 10 litres of water
- Production and distribution of heat to move turbines, to produce electricity
- Cleaning, irrigation of plants
- Preparation of food and drinks

Only the last two are related to domestic use. We will come back to this later.

Water as a standard

- Isotopes.
  There are 3 isotopes of hydrogen: ^2H, ^2D and ^2T, and three isotopes of oxygen, so theoretically there would be 9 forms of water. D_2O represents 0.02 %, the others even less. Tritium is radioactive.

- Water has its greatest density at 3.98°C. Ice therefore has a lower density and remains afloat; otherwise, the ice would sink, causing the bottom of the oceans to be pure ice: imagine the consequences for maritime life, for thermoregulation of the atmosphere.

- By definition, *1 litre* is the volume of 1 kilogram of water at 4°C under atmospheric pressure. The *specific gravity* of all substances is defined referring to water. The *Celsius scale* is made, referring to water: 0° being the equilibrium between water and ice under 1 atm. pressure, and 100°C being the equilibrium between water and steam, under 1 atm.

- The calorie is the amount of heat, needed to make 1 gram of water 1°C warmer. Or more exactly: 1/100 of the heat needed to warm 1 gram of water to 100°C. Compared to many other substances, water needs much energy to warm up. To put it another way: water **slows down the rise in temperature**; this buffer is extremely important for climate and biology. Just imagine that the heat, produced by a human in one day, would be able to increase his temperature by 150°C, if there were not this specific capacity of water.
  If water at 0°C is cooled, ice will form. But it takes 80 (negative!) calories per gram to make ice at 0°C. This latent heat plays an important role in the buffering of climate. Evaporation takes a lot of energy as well. In order to turn 1 gram of water at 100°C into 1 gram of steam at 100°C, 540 calories are needed. This enormous quantity of latent heat will be liberated during the reverse process, i.e. the condensation. For the third time, water demonstrates its important, vital role of thermo-regulation.

- 1 litre of water makes 1,700 litres of vapour. Seawater freezes at about -2°C.

- Compared to other substances with an analogous chemical structure, water has a high melting point and boiling point. In other words: these other substances are gaseous at these current temperatures. This property of water is essential for climate and biology.

- The structure of the water molecule is not rectilinear, but has the form of an isosceles triangle, with the O-atom in the top, the top angle measuring 105°C; in ice 6 molecules form an hexagon by means of hydrogen-bonds; in fluid water there are short chains of 2-5 molecules, in vapour all molecules are free. It is this change of configuration that may explain the high latent heat of melting and of evaporation.

- We saw that ice floats and expands. The pressure that ice exerts on pipes is many tons per cm². This means that even pipes of many centimetres thick will not resist, and eventually burst. They say, that keeping the water running, very slowly, drop by drop, could prevent the freezing.

- Living creatures consist mainly of water: humans about 70%, babies about 80%, an adult needs about 2-2.5 litres of water a day (drink and of food).

- It is very important to realize, that in nature and in biology, water is not consumed, but just borrowed; for the time being, it is not available to other purposes. But, since the beginning of time, about 4.5 billion years ago, the same water circulates and recycles constantly. There is no new water.

Water in the household

An average family uses about 140 litres of water per day and per person; yet the total consumption of a New Yorker is 1,000 litres per day!

For toilet	50 litre
Bath/ shower	30 litre
Dishes, laundry	35 litre
Cleaning	10 litre
Misc. (car, garden)	9 litre
Food, drinking	6 litre

Note: 1 bath takes about 200 litres, a shower about 20 litres per minute.
Note: only 6 litres a day must have "drinking-water quality" for the remainder rainwater could do.

This is a plea for more water tanks near new-built houses; it must be clear, that the cost of tap water will continue to rise: increasing cost of production, and increasing cost of disposal of the sewage.

We can easily do with less than 140 litres a day, without jeopardising our comfort.

The prevalence of water

Water abounds on Earth: 1,359,750,000 km³; but 97% of it is salty, only 3 % is fresh. Part of this is not available, (inlandsis, water in the atmosphere). Only the water in rivers and lakes and the ground water can be used. This means **0.6 % of the total!**

Snow offers the most pure water, as rain may contain more gasses (CO_2, etc.) but also more chlorides, sulphates, nitrates, ammoniac, as well as organic and inorganic dust.

Lakes and mountain rivers contain little organic material; they do contain salts. The same goes for wells, who may be polluted by human activities, e.g. by pesticides.

Even worse off are lowland rivers. Agriculture, mainly the over-fertilisation, and the input of sewage, should be forbidden in zones, where the ground water layer is to be exploited.

We definitely need that water reserve.

Water layers and oceans

- The presence of ground water may differ from region to region; but as an estimate, if one could pump up all the ground water of the USA, and spread it equally over the entire country, the water layer would be 33 m. thick.
- The oceans, with their salty water, may not be useful as drinking water, they still are extremely important for the climate. A few data to illustrate this:
    - The word "Ocean" is derived from the Greek *Okeanos*, the outer sea, in contrast to the better known inner seas.
    - The science of the oceans is called oceanography; limnology is about the inland water.
    - The ocean covers 70.8 % of the earth surface: 60.6 % in the northern hemisphere, and 81 % in the southern.
    - If the Earth would be equalised, and all the earth masses be dumped into the depth of the seas, then this equalised surface would be 8,000 feet (2,700 m.) under the actual sea level.
    - The mean depth is 4,150 m., the greatest depth (the Mariana Trench) is 12,000 m. deep. Yet all this constitutes only a thin film over the Earths surface. The Earth still is a perfect sphere, with a tolerance of less than 1 pro mille.
    - As a consequence of the Earth's rotation, of the sun's heat, and of the predominant winds, several streams are formed, e.g. the Gulfstream; near the equator streaming from East to West, and near the poles from West to East. This circling movement is completed alongside the continents.
    - The water masses turn clock-wise in the Northern hemisphere, and counter-clock wise in the southern. These water movements are extremely important for the homeostasis of the climate; they constitute a living proof of the buffering effect of water.
    - For the same reasons, the air masses follow the same pattern, but they are not hindered by the shores. Here too, evaporation (= energy uptake) over the oceans, is followed by precipitation (energy delivery) elsewhere, thus assuring a buffering and stabilising and equalising effect on the climate. When the Earth tends to warm up, the temperature will at first not rise; but more ice will melt, the currents in the sea will be more intense, the tropical storms will appear further from the Equator.

Whoever hopes that global warming is welcome because it may save heating oil, may well be confronted with some nasty side effects.

The water cycle

It is well known that water evaporates from the ocean, but plants contribute greatly to this phenomenon. One full-grown corn plant needs about one gallon a day. One acre (4,046 m²) of corn needs 15,000 litres of water per season, i.e. about 40,000 litres per hectare. One big oak needs 400 litres a day.

Part of this water is to build up the plant's tissue via photosynthesis. But a major part just evaporates. Here again, we see the strong thermo-regulating effect of water: plants stay cooler, they provide air humidity, and when the planted area is big enough, they may cause self-sustaining rainfall, e.g. in the tropical rain forests.

Global warming is not just the result of greenhouse gasses. Desertification and destruction of the tropical forests, and the presence of megacities full of stone, concrete and asphalt that create their own microclimate, are obvious causes.

Water evaporates from the sea, perspires from plants by sun heat, wind takes care of the transport, colder air causes condensation, precipitation is a consequence of gravity. The reflux of water to the sea, eventually via a bypass via ground water, also is the consequence of gravity.

Of the rain, about 2/3 evaporates directly or via the plants; 1/3 reaches brooks and rivers, and ultimately the sea, or the deeper ground layers. A source may bring it to surface again, so it will still end in the sea.

The 0.6 % of water, that are at disposal of men and animals and plants, are not statically present: they are available only for a short period, during their cycle from the sea, and back to the sea; the faster that cycle, the shorter the time we can use that water. This cycle is accelerated by chopping of mountain slopes, by canalisation of rivers, and by constructions (roads, houses, industry) who diminish the surface, where rainwater can seep into the underground, diminish the phreatic water, and accelerate the flow-off of water, with higher risk of flooding.

Since ages, man has tried to **ensure his water supply**.

The first "modern" water works, with pump and reservoir, were built in London, in 1582; the wooden pipes allowed low pressure only. Not before 1800 iron pipes were used.

In the USA the first installation was built in Boston in 1652; by 1800, 16 major cities disposed of a water works system.

But as soon as 4000 BC, in Mesopotamia, open canals were used to irrigate fields. The first draw-wells were active in Mohenjo Daro (now: Pakistan) in 3000 BC. The first pumps were the *Shadoof*, in 1600 BC in Egypt, man-moved. The system is still in operation in Egypt, the Middle-East and India. The *Persian Wheel* originated in 200 BC - the "moving force" here was an animal. The Persian Wheel (Saqiya) is still in use. Not before 1700 AD, steam-engines and pumps were used.

In the 6th century BC, in Samos, a tunnel was made through a mountain, in order to ensure water supply to a fortified city, so that it could better resist enemy sieges.

Rome had 11 aqueducts, the first as soon as 312 BC. There was enough water, even for parks and fountains.

During the middle ages, the lack of infrastructure led to massive, mortal epidemics: typhoid, cholera and dysentery.

The first desalination plant started in 1949 in Kuwait. Now, in 1996, over 2000 are operational worldwide, half of them in the Middle East. Their action is based on one of three principles: distillation, evaporation under low pressure, or reverse osmosis.
They need too much energy to be remunerative world-over.
The first chemical purification of water was performed in 1908, using chlorine.
The largest modern aqueduct starts from the Colorado River (USA) and measures 380 km. It provides water to San Diego and Los Angeles.

Conclusion

Water has several unique physical properties. It is liquid at "physiological" temperature, it has a high heat capacity, it has high melting– and evaporation warmth, its density decreases below 0°C, et cetera.

If any one of these properties were not present, then life as we know it would be impossible.
Water is not "used"; the same water recycles for 4.5 billion years, over and over again.
During that cycle is at our disposal: just 0.6 % of the total amount on Earth!

If we want to continue to use water, then we must slow down the cycle and stop water pollution, because that would make water temporarily unavailable and more expensive still.
This speech tried to propose a few suggestions on this item.

But all our efforts are vain, as long as number one problem is not solved: **overpopulation.**
I hope to have found a forum here to tackle this problem.

Dendermonde, January 4, 1997.

The choice is ours:

Either

an efficient policy of birth control
PLUS a few measures on immigration,
desertification, reforestation and fossil fuels.

Or

no efficient birth control, but
depletion of all reserves
PLUS famine
PLUS depletion of the fishing stocks
PLUS water shortage
PLUS global warming
PLUS more desertification
PLUS more deforestation
PLUS problems about CO^2
PLUS overpopulation
PLUS more slums
PLUS on-going emigration pressure
PLUS more genocide

Dawn is over